Singapore

Grade 2

We try to encourage students to read and think about the problems carefully. We facilitate this by:

ACTIVITIES

You and your child will discover that our second grade math activities are designed to strengthen newly learned concepts and reinforce old ones.

WORD PROBLEMS

We provide math word problems for addition, subtraction, time and fractions.

MATH GAMES & CHALLENGES

Great games and challenges that will sharpen the brain and focus.

Table of Content

Geometry

Identifying Shapes

Write the name of each shape using the following words.

Rhombus Sphere Cercle Triangle Rectangle Square
Trapezoid Parallelogram Hexagon Oval Cone Pentagon

Geometry

Identifying Shapes

Write the name of each shape using the following words.

Trapezoid Pyramide Hexagonal Prism Oval Cone

Cuboid Sphere Cercle Cube Cylinder Star Pentagon

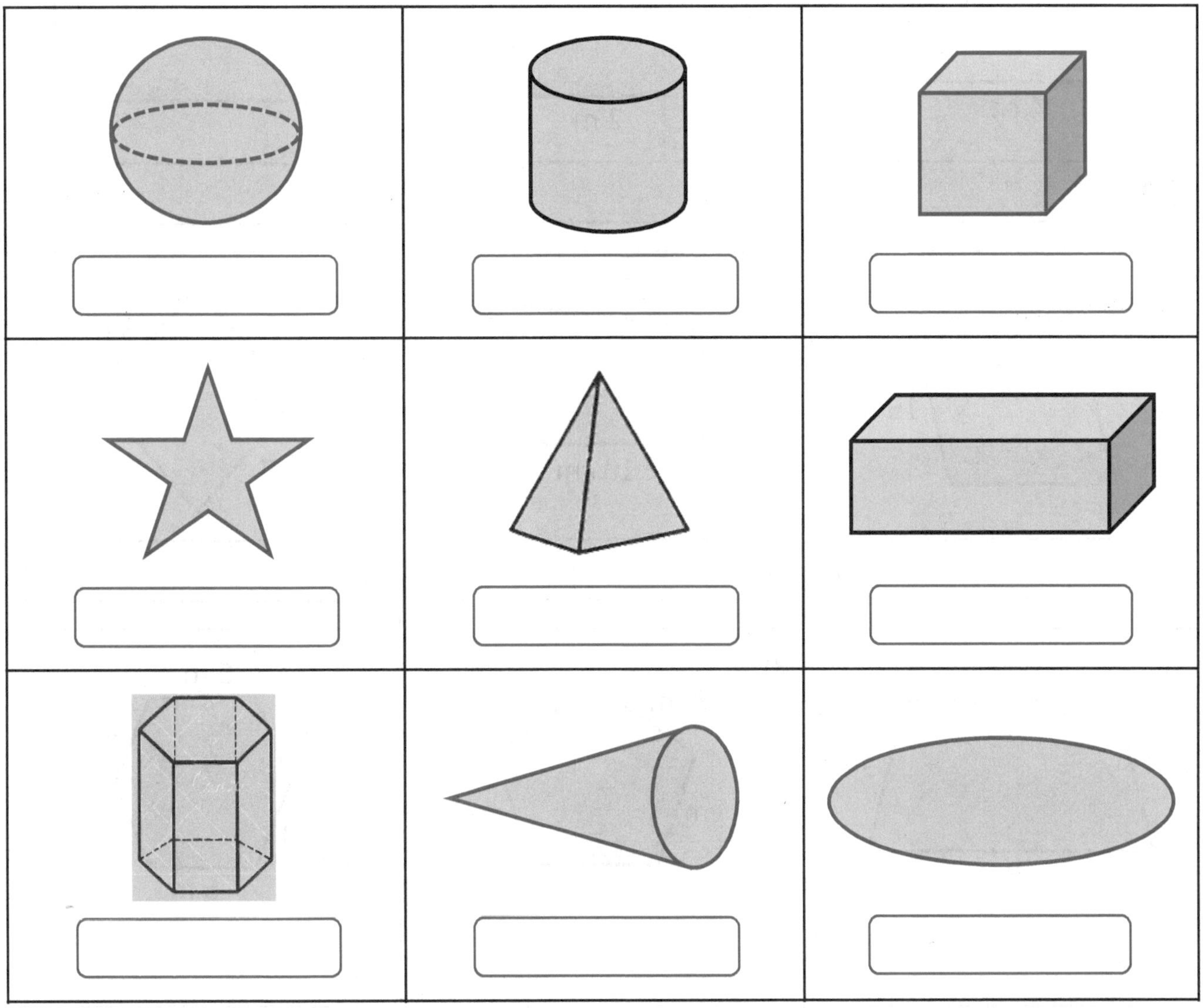

Geometry

Perimeters of 2D Shapes

Find the perimeter of the shapes shown below.

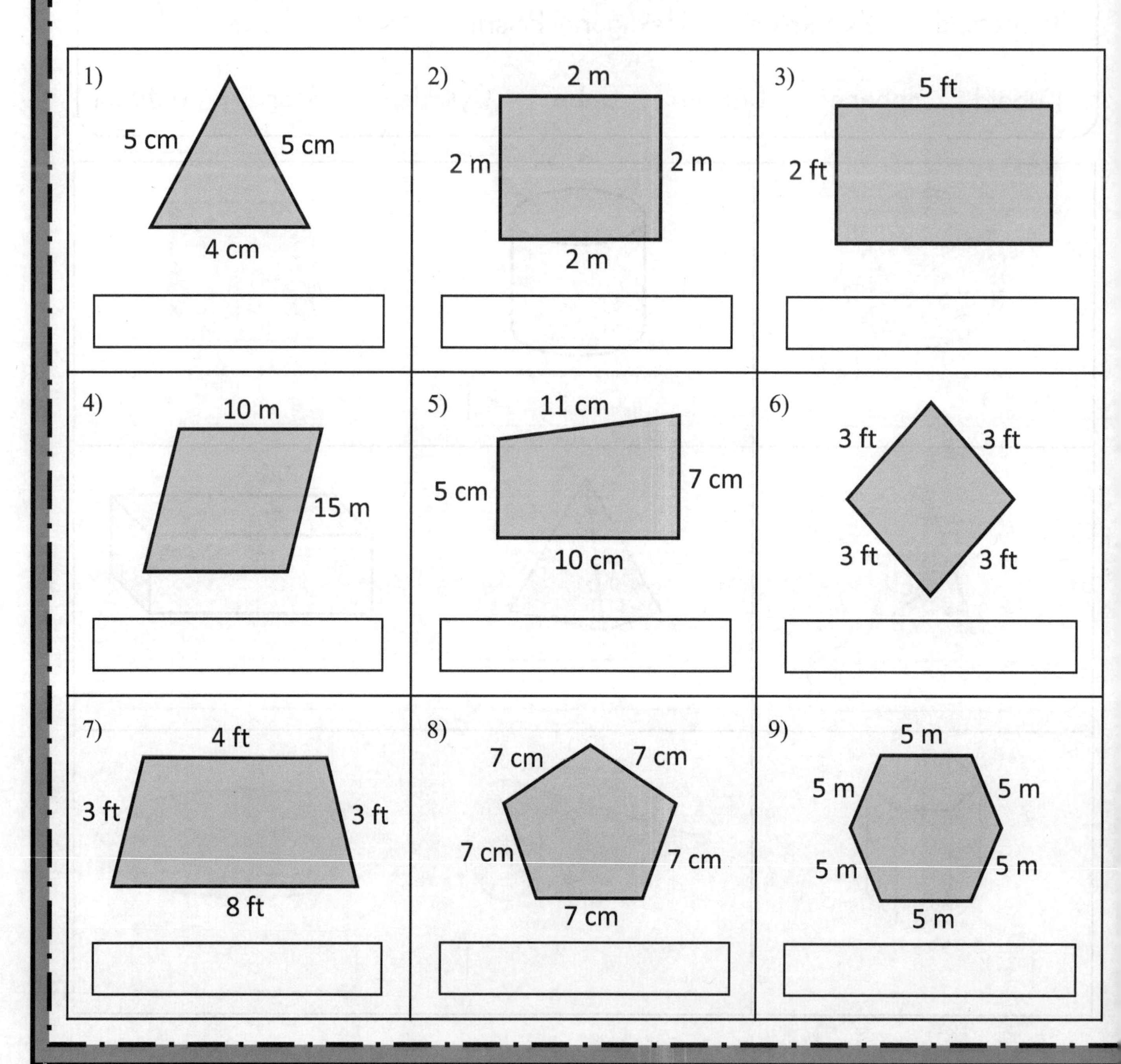

Geometry

Sides and Vertices

Fill in the following table.

Shape	Number of Sides	Number of Vertices

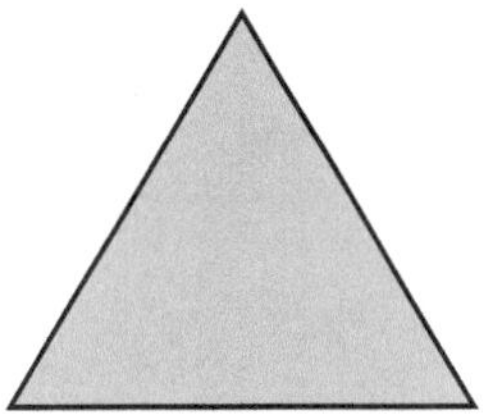		
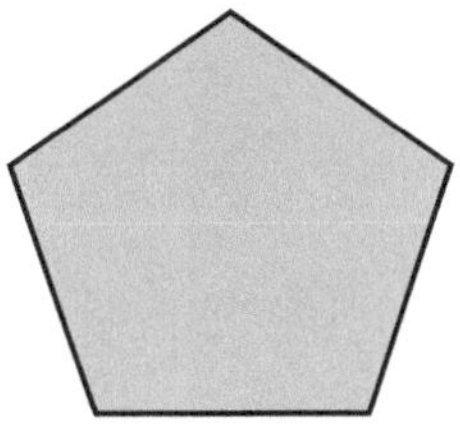		

Geometry

Sides and Vertices

Fill in the following table.

Shape	Number of Sides	Number of Vertices
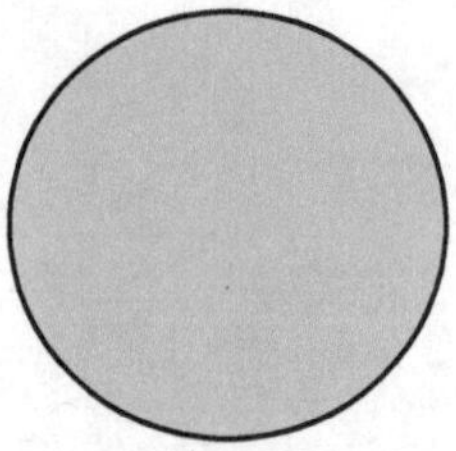		

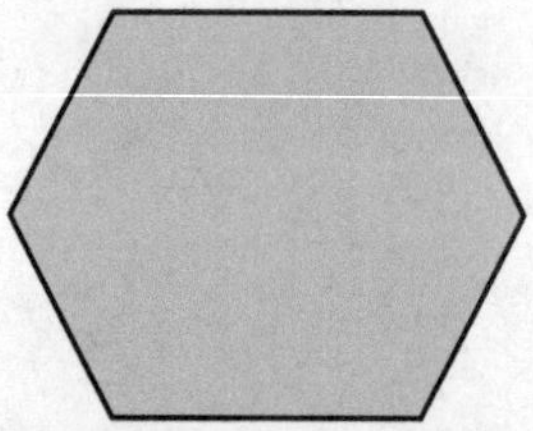		

Geometry

Faces, Edges & Vertices

Fill in the following table.

Shape	Number of Faces	Number of Edges	Number of Vertices
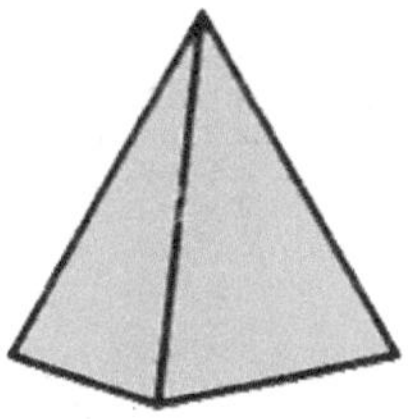			
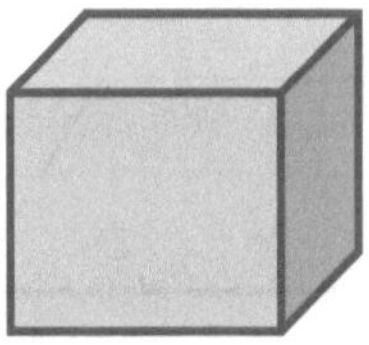			
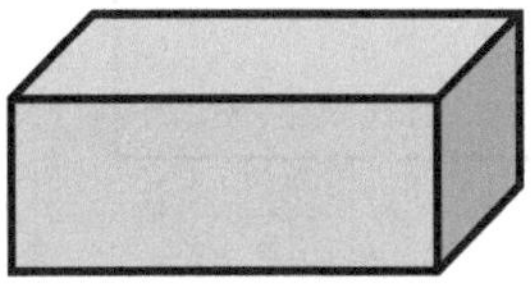			
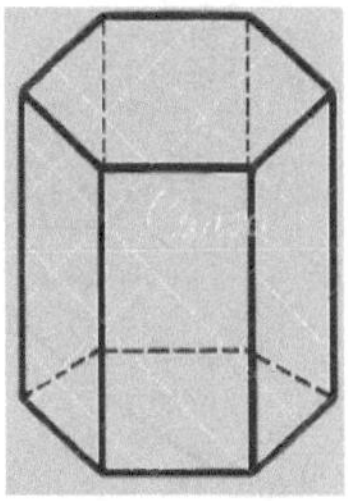			

Lines of Symmetry

Draw a line that cuts the following shapes in half, so that each half reflects the other half through your line.
Hint: Some shapes can be cut in more than one way.

Geometry

Symmetry of 2D Shapes

Drawing the other half of the following symmetric shapes.

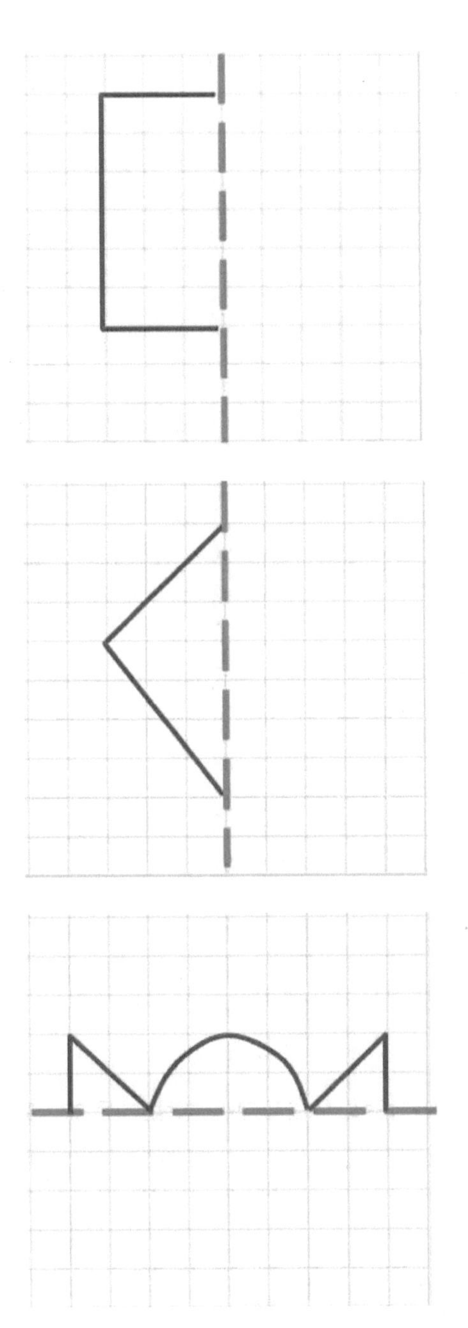

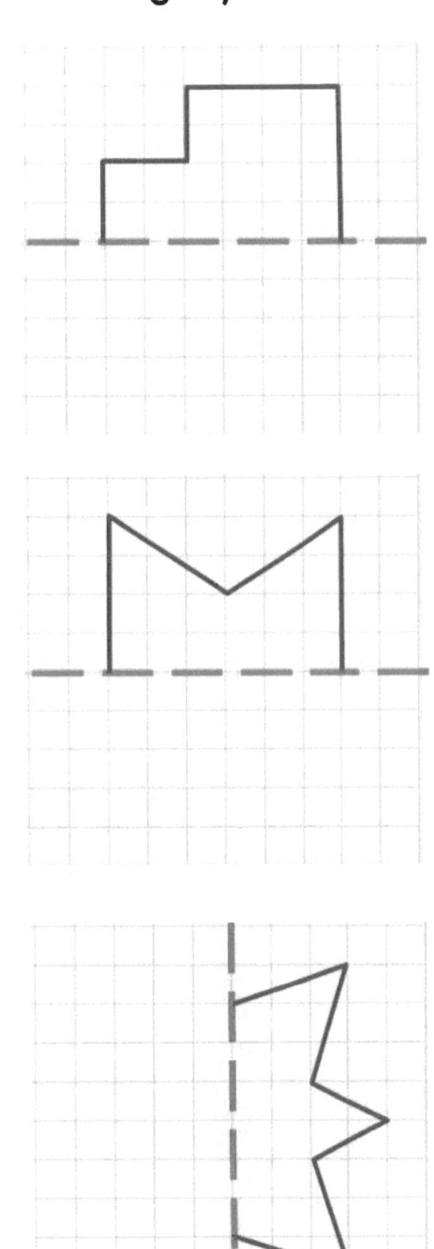

Geometry

Symmetry of 2D Shapes

Drawing the other half of the following symmetric shapes.

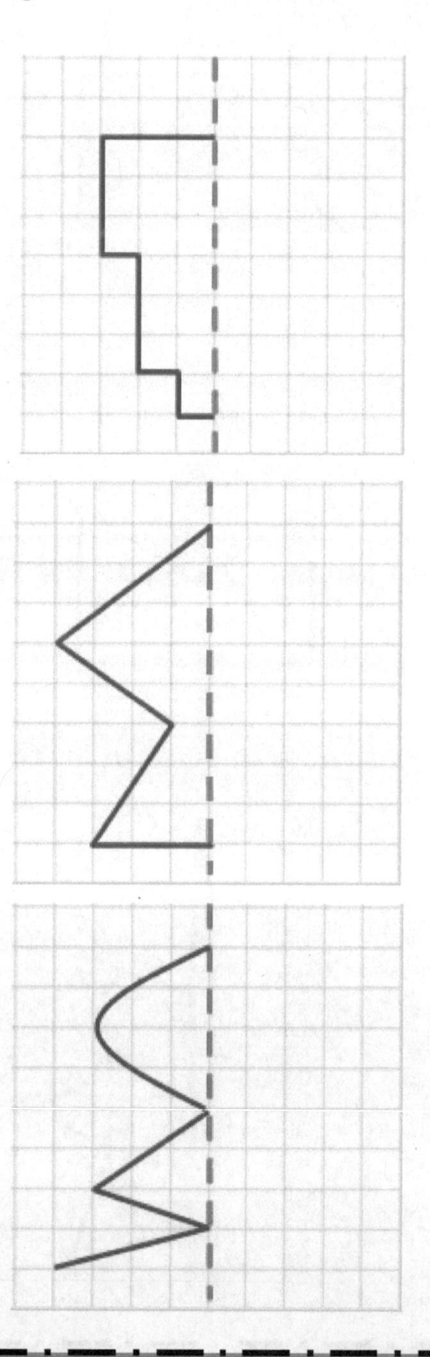

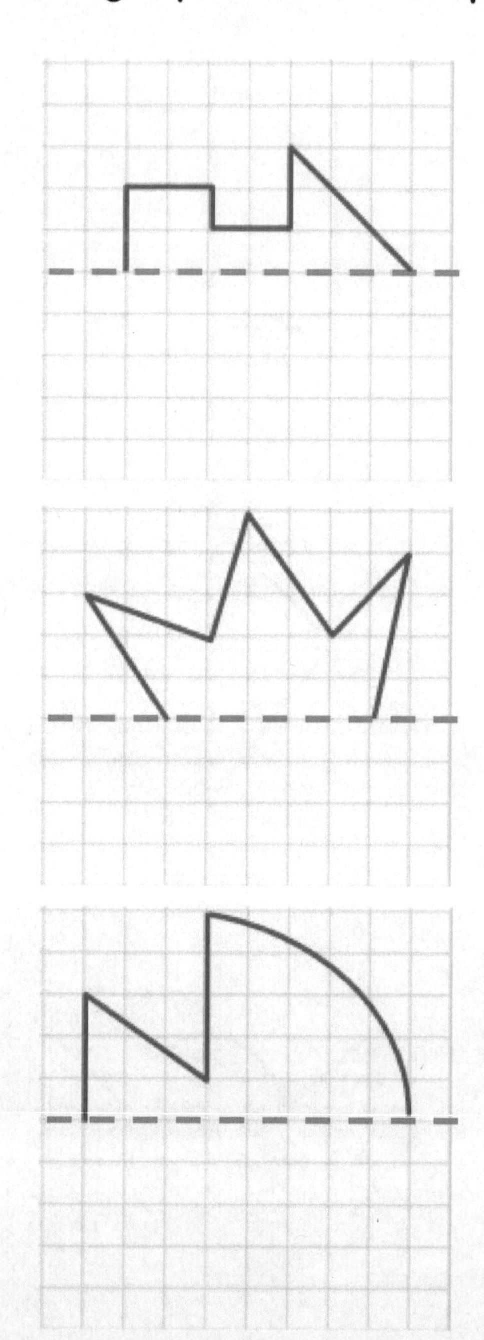

Geometry

Game and Challenge

Drawing the other half of this symmetric rocket.

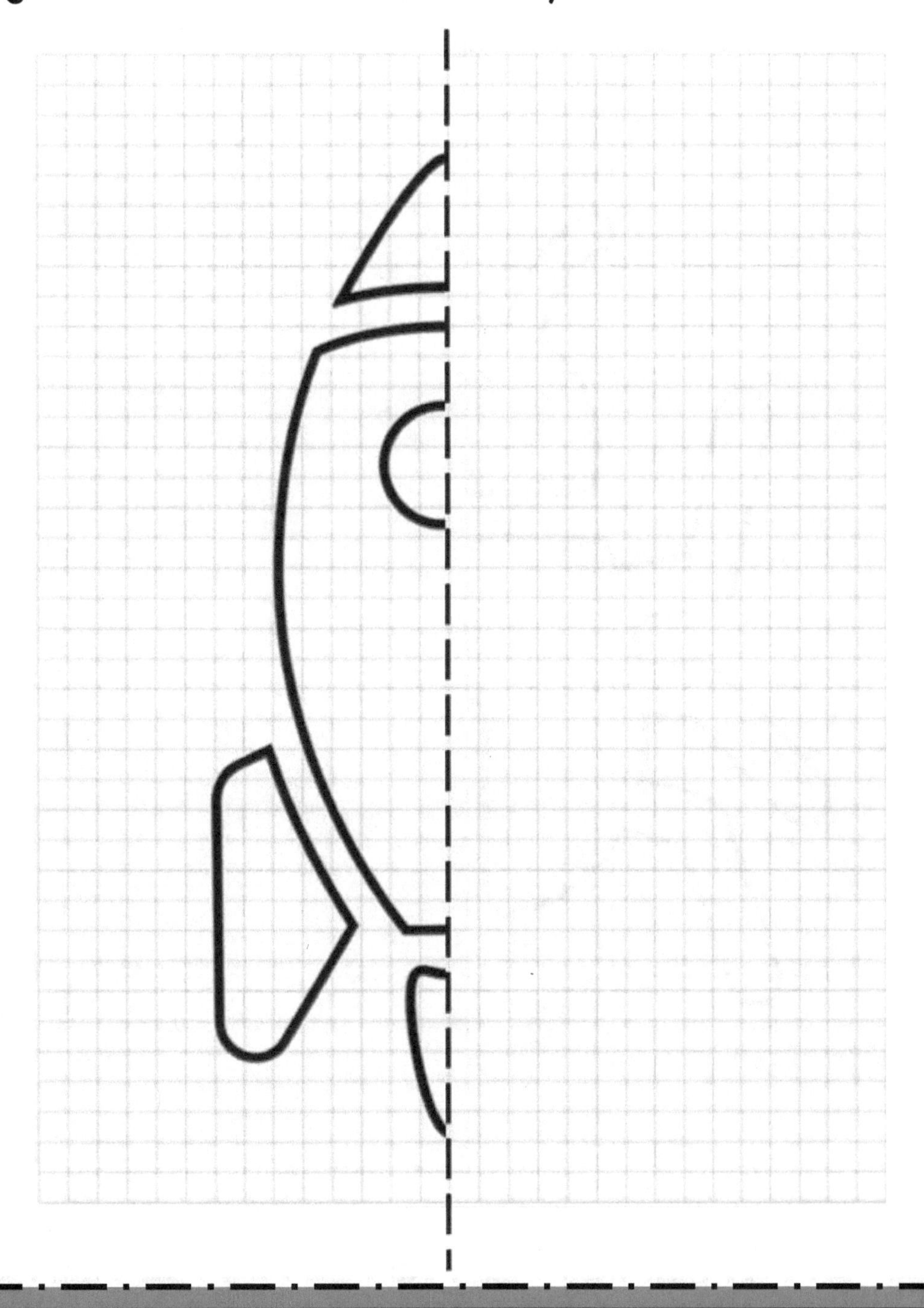

Game and Challenge

Drawing the other half of this symmetric bug.

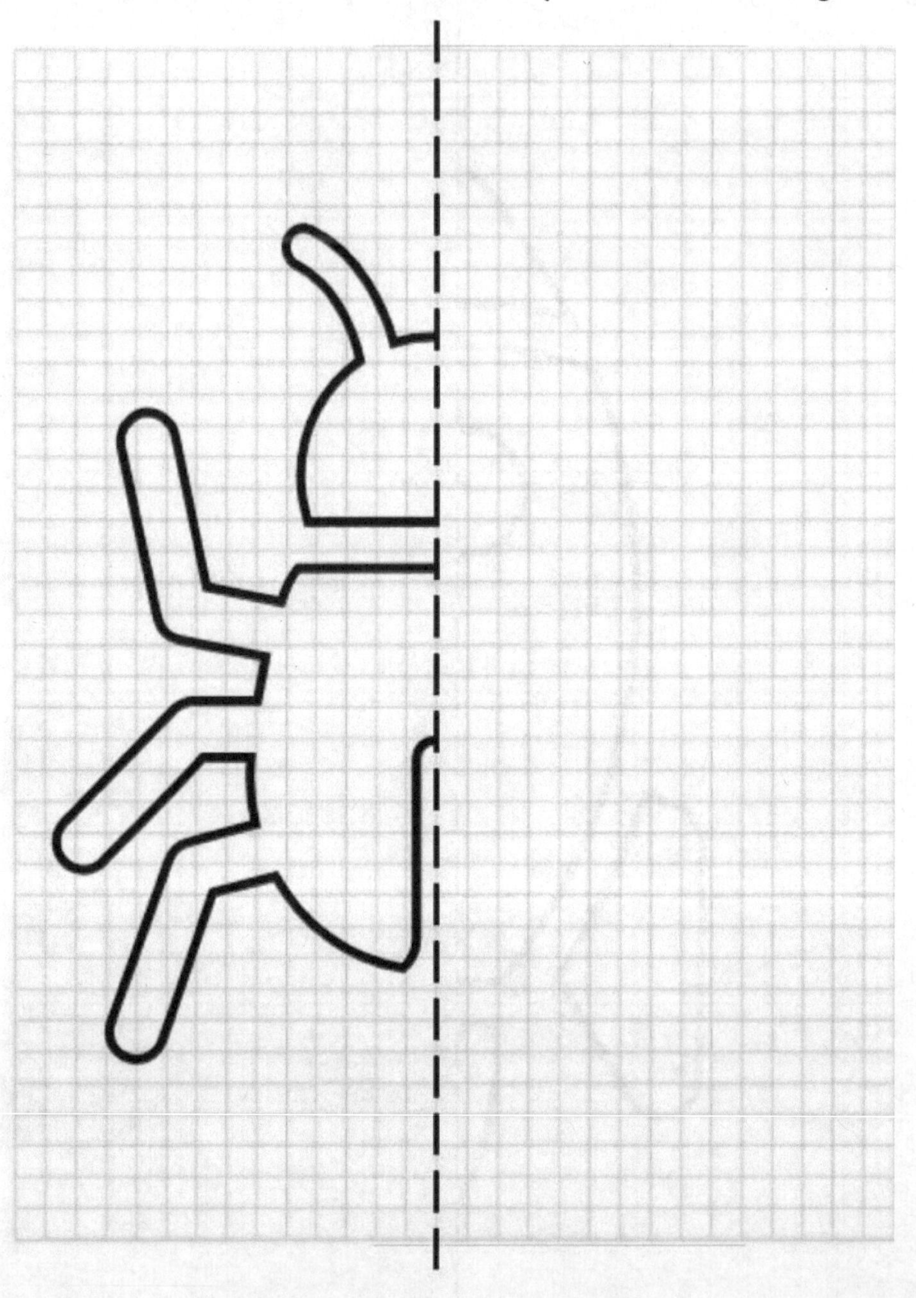

Adding in Columns

Adding 2-digit and 1-digit numbers (no regrouping)

Find the sum.

$$\begin{array}{r} 36 \\ +\ \ 3 \\ \hline \end{array} \quad \begin{array}{r} 24 \\ +\ \ 1 \\ \hline \end{array} \quad \begin{array}{r} 52 \\ +\ \ 4 \\ \hline \end{array} \quad \begin{array}{r} 40 \\ +\ \ 2 \\ \hline \end{array}$$

$$\begin{array}{r} 23 \\ +\ \ 4 \\ \hline \end{array} \quad \begin{array}{r} 13 \\ +\ \ 5 \\ \hline \end{array} \quad \begin{array}{r} 70 \\ +\ \ 1 \\ \hline \end{array} \quad \begin{array}{r} 65 \\ +\ \ 4 \\ \hline \end{array}$$

$$\begin{array}{r} 82 \\ +\ \ 5 \\ \hline \end{array} \quad \begin{array}{r} 71 \\ +\ \ 8 \\ \hline \end{array} \quad \begin{array}{r} 47 \\ +\ \ 1 \\ \hline \end{array} \quad \begin{array}{r} 93 \\ +\ \ 3 \\ \hline \end{array}$$

$$\begin{array}{r} 88 \\ +\ \ 0 \\ \hline \end{array} \quad \begin{array}{r} 91 \\ +\ \ 1 \\ \hline \end{array} \quad \begin{array}{r} 60 \\ +\ \ 0 \\ \hline \end{array} \quad \begin{array}{r} 92 \\ +\ \ 7 \\ \hline \end{array}$$

Adding in Columns

Adding 2-digit and 1-digit numbers (with regrouping)

Find the sum.

$$\begin{array}{r} 24 \\ +\ 7 \\ \hline \end{array} \quad \begin{array}{r} 38 \\ +\ 6 \\ \hline \end{array} \quad \begin{array}{r} 17 \\ +\ 5 \\ \hline \end{array} \quad \begin{array}{r} 66 \\ +\ 4 \\ \hline \end{array}$$

$$\begin{array}{r} 73 \\ +\ 9 \\ \hline \end{array} \quad \begin{array}{r} 55 \\ +\ 8 \\ \hline \end{array} \quad \begin{array}{r} 41 \\ +\ 9 \\ \hline \end{array} \quad \begin{array}{r} 88 \\ +\ 8 \\ \hline \end{array}$$

$$\begin{array}{r} 57 \\ +\ 7 \\ \hline \end{array} \quad \begin{array}{r} 29 \\ +\ 8 \\ \hline \end{array} \quad \begin{array}{r} 66 \\ +\ 9 \\ \hline \end{array} \quad \begin{array}{r} 45 \\ +\ 6 \\ \hline \end{array}$$

$$\begin{array}{r} 73 \\ +\ 9 \\ \hline \end{array} \quad \begin{array}{r} 86 \\ +\ 6 \\ \hline \end{array} \quad \begin{array}{r} 79 \\ +\ 9 \\ \hline \end{array} \quad \begin{array}{r} 77 \\ +\ 9 \\ \hline \end{array}$$

Adding in Columns

Add two 2-digit numbers in columns (no regrouping)

Find the sum.

53 + 21 = ____	31 + 40 = ____	15 + 12 = ____	67 + 11 = ____
26 + 70 = ____	72 + 17 = ____	44 + 31 = ____	88 + 10 = ____
55 + 34 = ____	61 + 31 = ____	33 + 33 = ____	72 + 15 = ____
50 + 40 = ____	67 + 20 = ____	40 + 30 = ____	80 + 19 = ____

Adding in Columns

Add two 2-digit numbers in columns (with regrouping)

Find the sum.

64 + 36	67 + 46	45 + 79	55 + 17
28 + 36	45 + 45	37 + 66	14 + 59
64 + 18	38 + 79	87 + 67	59 + 97
88 + 37	46 + 99	68 + 78	19 + 89

Adding in Columns

Add three 2-digit numbers in columns

Find the sum.

60 35 + 14	38 45 + 54	26 37 + 44	35 29 + 20
30 57 + 84	86 79 + 25	47 53 + 71	82 19 + 54
30 75 + 93	97 62 + 18	49 51 + 87	16 98 + 70
64 76 + 88	67 59 + 38	77 68 + 99	68 37 + 43

Adding in columns - Missing addend

Fill in the missing number.

56	47	39	68
+ ___	+ ___	+ ___	+ ___
64	54	43	73
78	85	17	77
+ ___	+ ___	+ ___	+ ___
87	88	25	80
94	85	72	34
+ ___	+ ___	+ ___	+ ___
95	85	77	43
45	68	90	22
+ ___	+ ___	+ ___	+ ___
49	70	95	29

Adding in Columns

Adding in columns - Missing addend

Fill in the missing number.

45 + ___ = 67	36 + ___ = 82	28 + ___ = 53	79 + ___ = 96
67 + ___ = 94	74 + ___ = 86	37 + ___ = 94	24 + ___ = 79
66 + ___ = 99	33 + ___ = 77	40 + ___ = 61	16 + ___ = 83
13 + ___ = 94	25 + ___ = 80	31 + ___ = 77	50 + ___ = 90

Adding in Columns

Adding 3-digit numbers in columns (no regrouping)

Find the sum.

```
  121      311      500      161
+ 163    + 248    + 358    + 821
-----    -----    -----    -----

  217      752      467      710
+ 650    + 143    + 532    + 211
-----    -----    -----    -----

  832      462      333      781
+ 106    + 312    + 521    + 200
-----    -----    -----    -----

  822      653      431      117
+ 105    + 315    + 566    + 421
-----    -----    -----    -----
```

Adding in Columns

Adding 3-digit numbers in columns (with regrouping)

Find the sum.

867 + 417	672 + 348	561 + 509	468 + 718
791 + 168	827 + 236	465 + 958	538 + 697
682 + 499	317 + 729	264 + 178	176 + 256
918 + 874	784 + 946	879 + 849	693 + 990

Adding three 3-digit numbers in columns

Find the sum.

208 346 + 573	643 218 + 374	259 743 + 883	840 881 + 122
795 815 + 369	815 746 + 937	538 867 + 976	857 674 + 190
228 999 + 112	438 595 + 631	801 919 + 476	549 276 + 638
963 857 + 784	891 702 + 380	157 679 + 401	786 527 + 941

Addition

The numbers in the circles added together makes the number in the linking rectangle. Find the missing numbers in this puzzle.

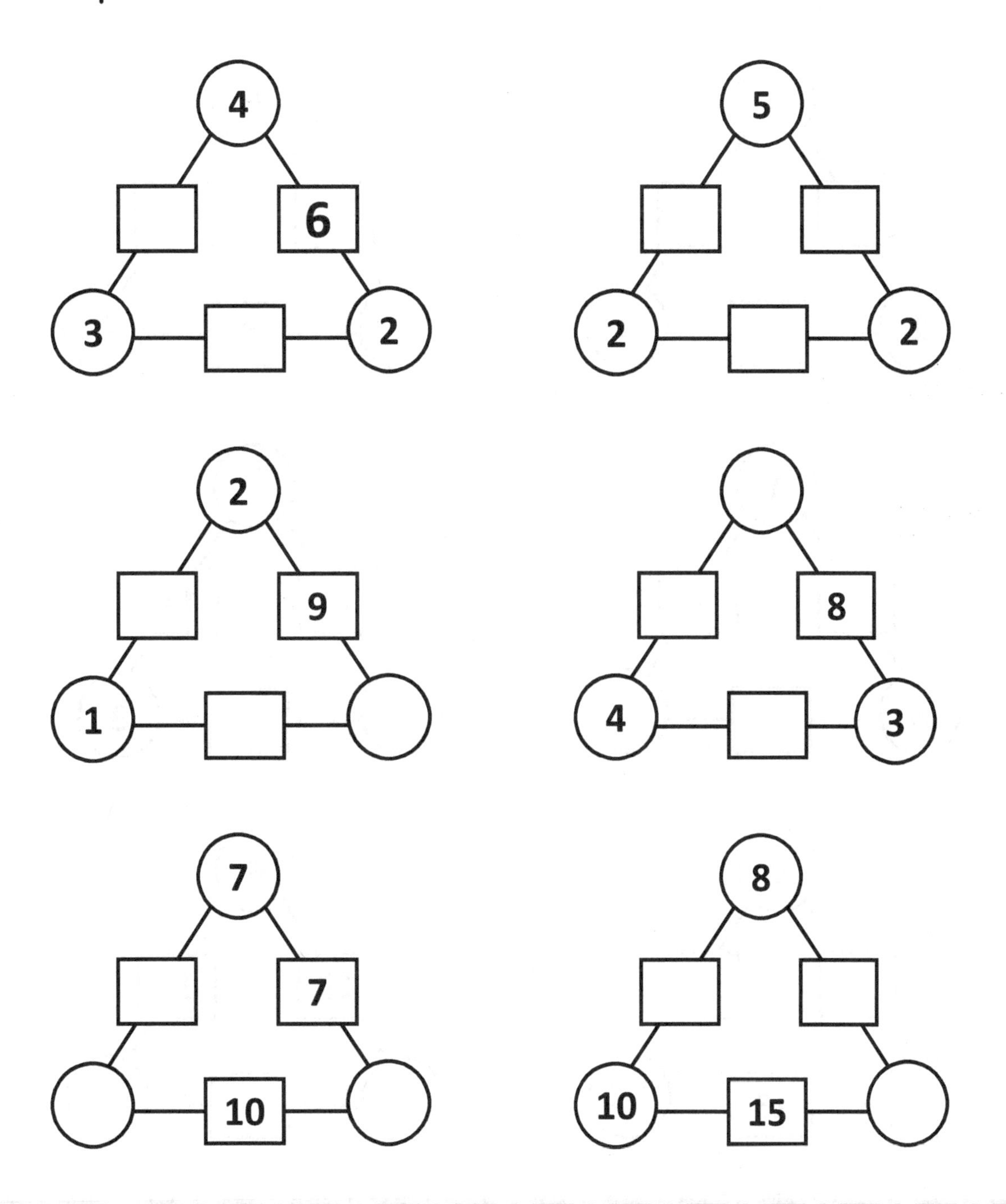

Addition

The numbers in the circles added together makes the number in the linking rectangle. Find the missing numbers in this puzzle.

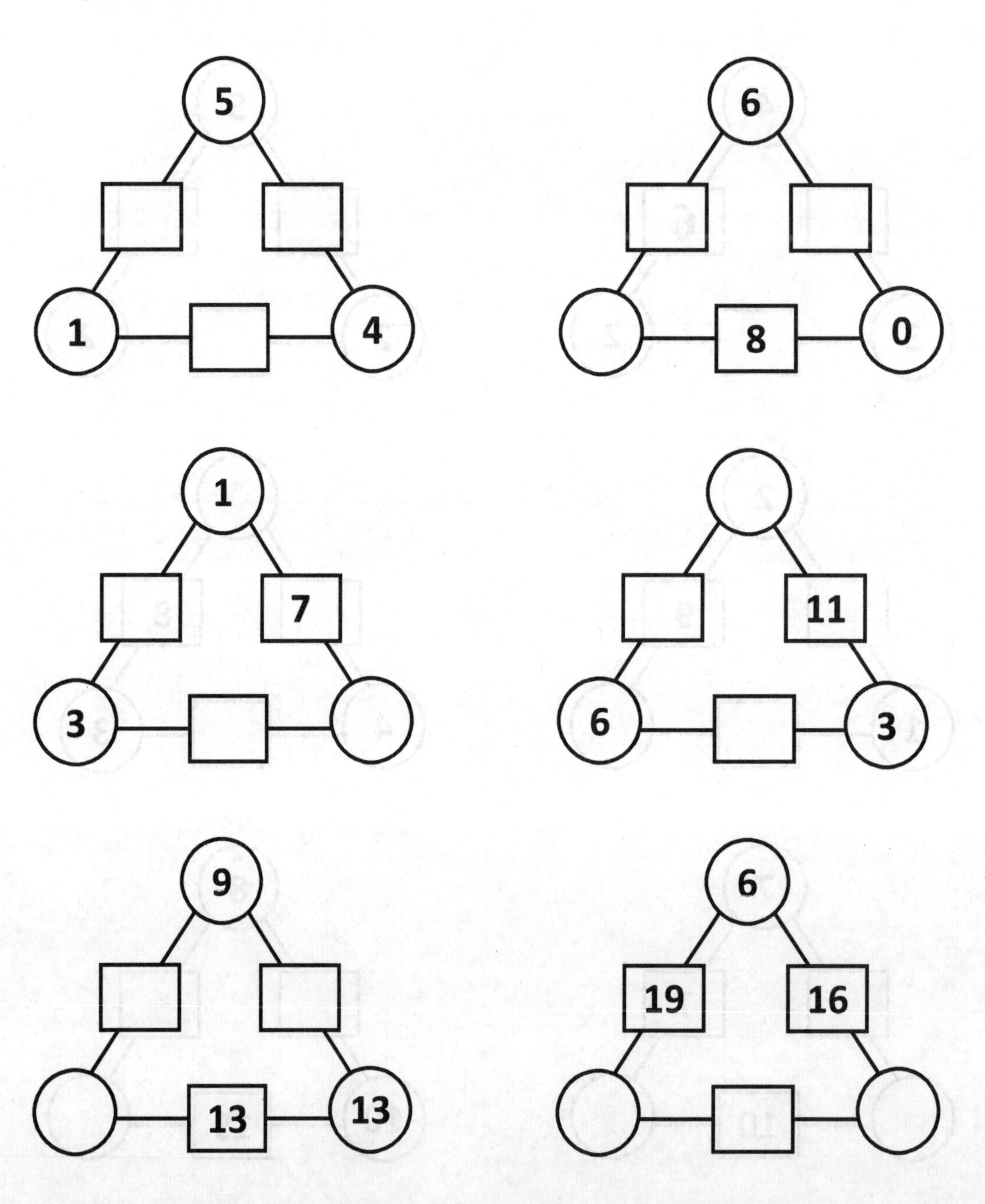

Addition

The numbers in the circles added together makes the number in the linking rectangle. Find the missing numbers in this puzzle.

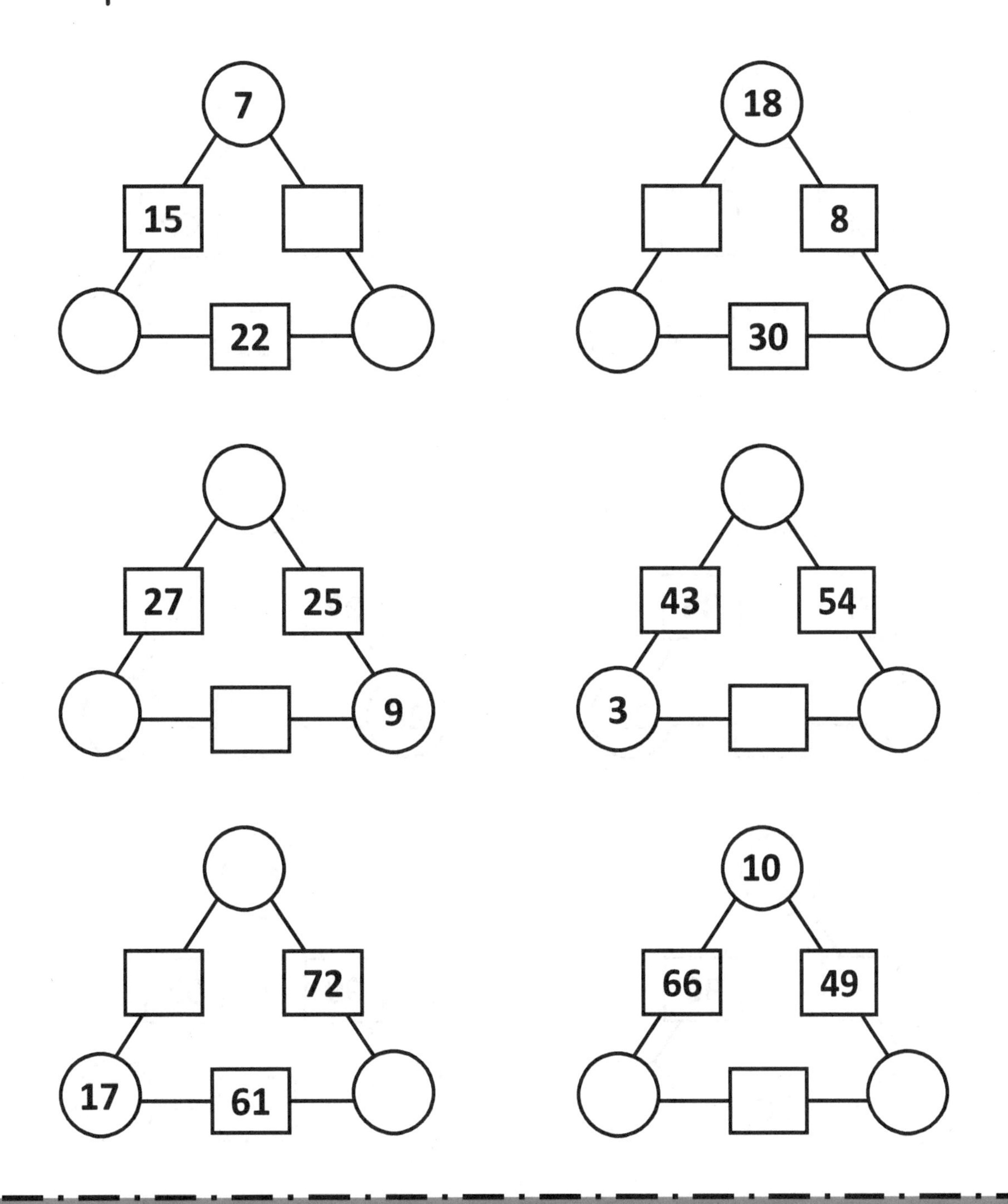

Addition

The numbers in the circles added together makes the number in the linking rectangle. Find the missing numbers in this puzzle.

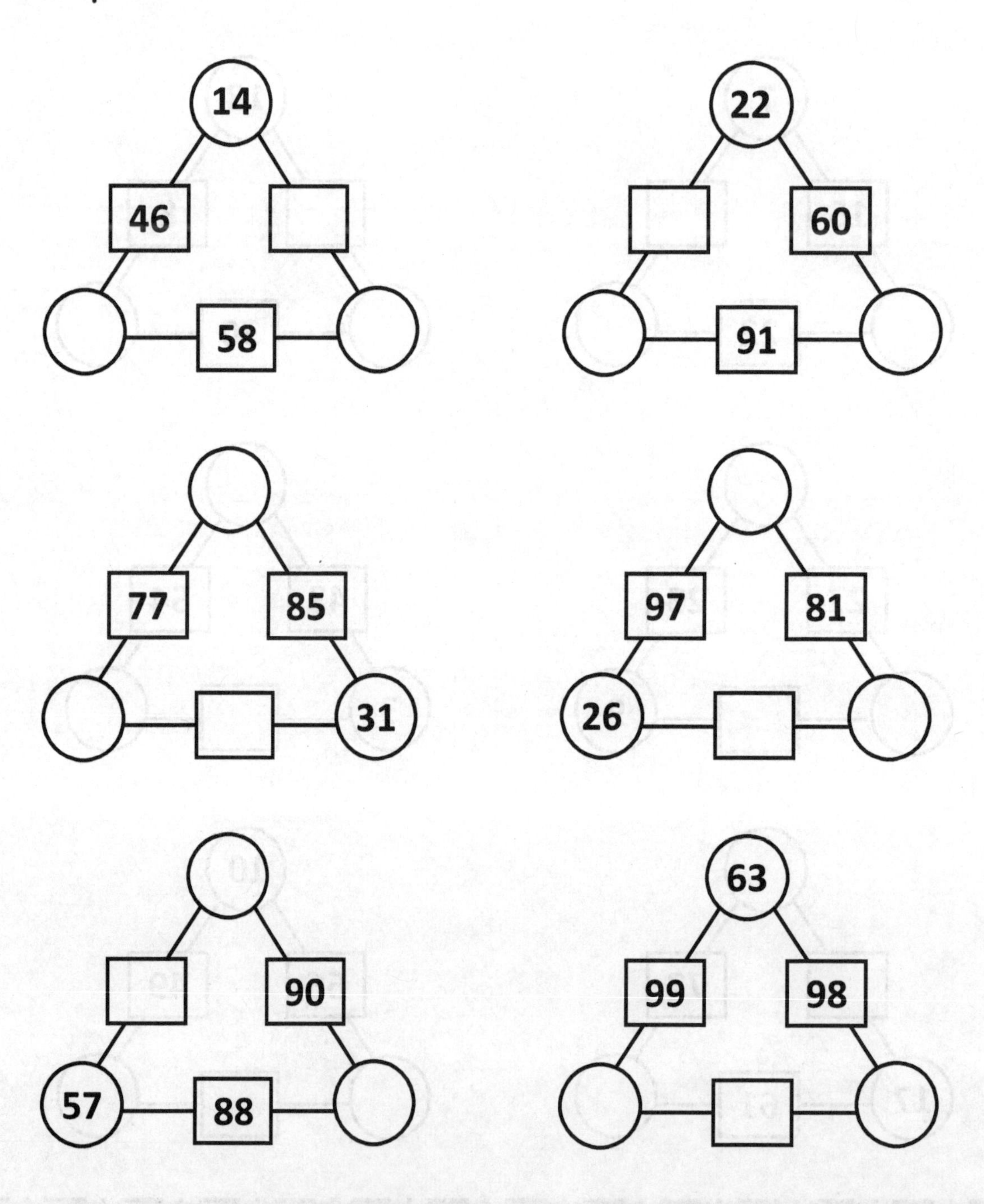

Addition

Word Problems

Read and solve the problems.

1) Catherine bought 2 candies for 7 cents and 3 bubble gums for 12 cents each. How much did she spend in all?

2) Sara made 7 rice krispie treats. She used 4 large marshmallows and 10 mini marshmallows. How many marshmallows did she use altogether?

3) Lucy ate 2 apples every hour. How many apples had she eaten at the end of 3 hours?

Addition

Word Problems

Read and solve the problems.

1) Emma saw four bugs eat five flowers each. How many flowers total did the bugs eat?

2) Isabella bought 3 pizzas. Each pizza had 9 slices. How many total slices of pizza did she have?

3) Benjamin read 2 books per day. How many books did he read in one week?

Addition

Word Problems

Read and solve the problems.

1) The local food bank opened 25 days in January for 10 hours a day and 27 days in February for 9 hours a day. If it will be opened for 30 days in March for 12 hours a day, how many days is it open in these three months?

2) The zoo is separated into 4 zones. The Artic zone has 6 exhibits. The African zone has 13 exhibits. The rainforest zone has 10 exhibits and the tropical zone has 20 exhibits. How many exhibits are there in the zoo?

3) Linda's hair is 20 inches long. If her hair grows 5 more inches each month, how long will it be in four more months?

Addition

Word Problems

Read and solve the problems.

1) Ashley, Bob and Clara are keeping score of the game they are playing. When a player wins a game, that player gets 5 points. If a player loses a game, the player has 3 points taken away. If it is a tie, every player gets 2 points. Each of them has 20 points to start with. Ashley wins the first game. How many points does Ashley has after the first game?

2) Amelia gave 4 pieces of candy to each student in the group. The group had a total of 10 students in it. How many pieces of candy did Amelia give away?

3) Each CD rack holds 20 CDs. A shelf can hold four racks. How many total CDs can fit on the shelf?

Addition

Word Problems

Read and solve the problems.

1) Lucy is buying treats for her 4 cats. If she wants to buy them 3 heart biscuits each, how many biscuits does she need to buy?

2) Mr. Jackson has written three books. The first book contains 365 pages, the second contains 781 pages, and the third contains 891 pages. How many pages are written by Mr. Jackson?

3) Sara filled her bucket with 125 pounds of shells. If she adds 119 more pounds of shell to fill her bucket, how many pounds does she have? Do you think she can pick up the bucket?

Addition : Game and Challenge

Use your math skills to find the value of each "?".

+ + = 5

+ = 2

= ?

= ?

+ = 3

+ + = 3

= ?

= ?

Addition : Game and Challenge

Use your math skills to find the value of each "?".

+ + = 6

+ = 4

+ = 6

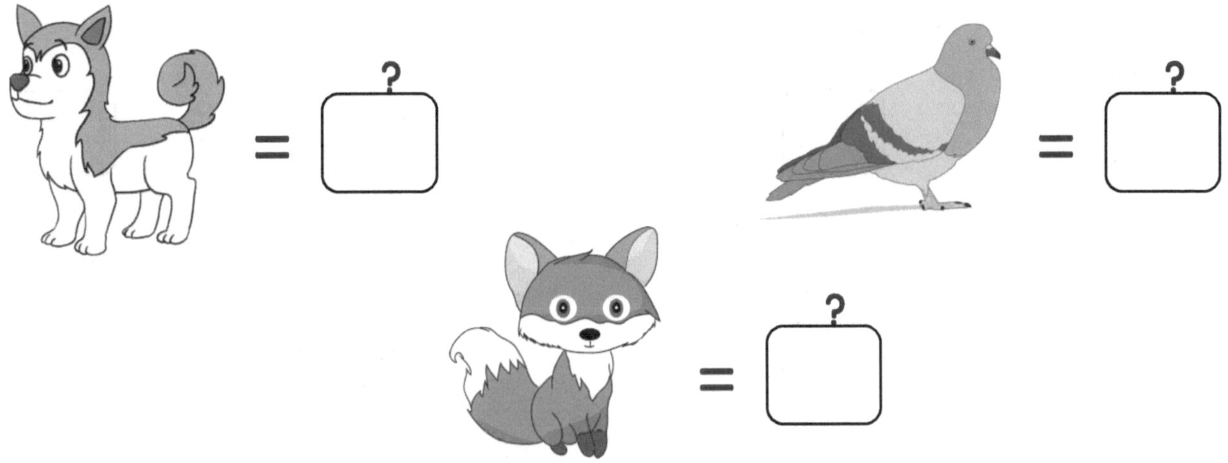

= ?

= ?

= ?

Addition : Game and Challenge

Use your math skills to find the value of each "?".

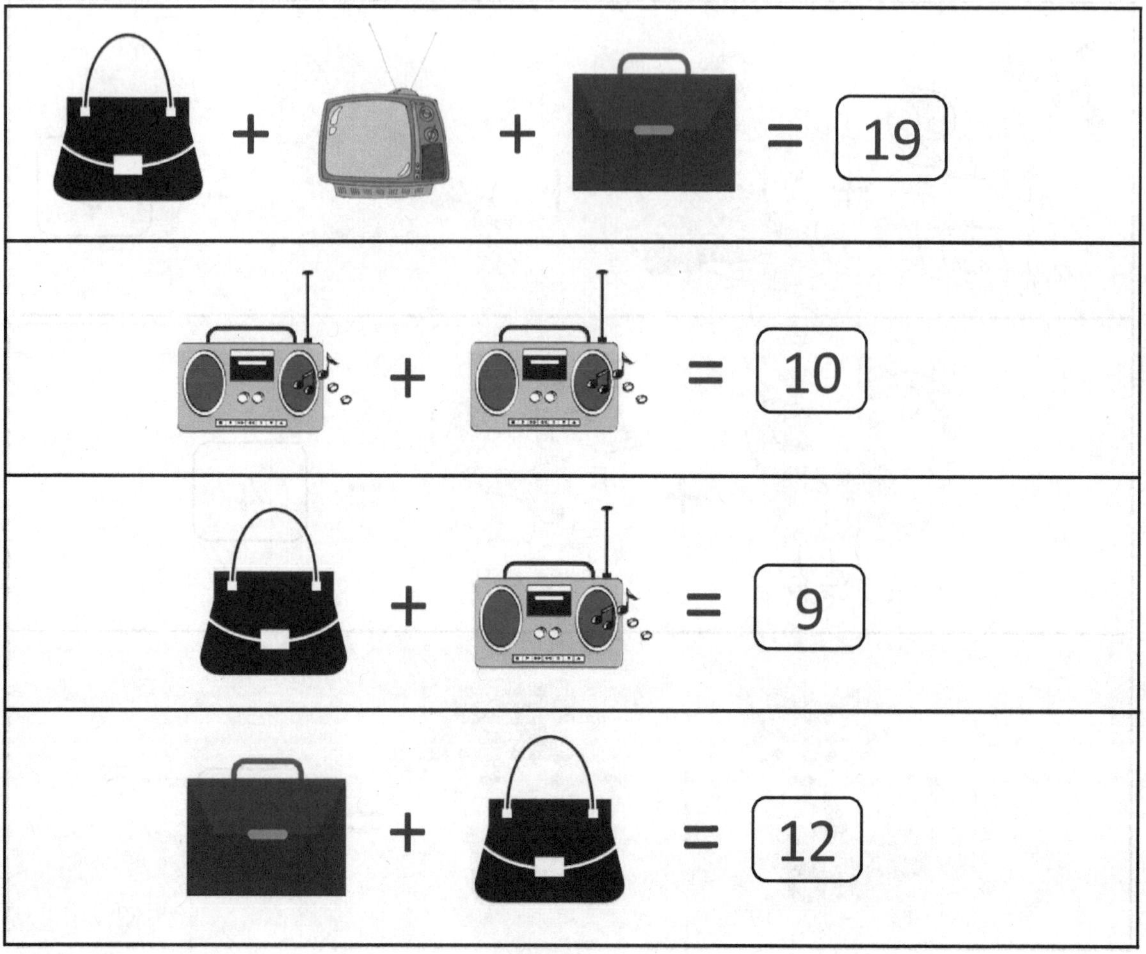

+ + = 19

+ = 10

+ = 9

+ = 12

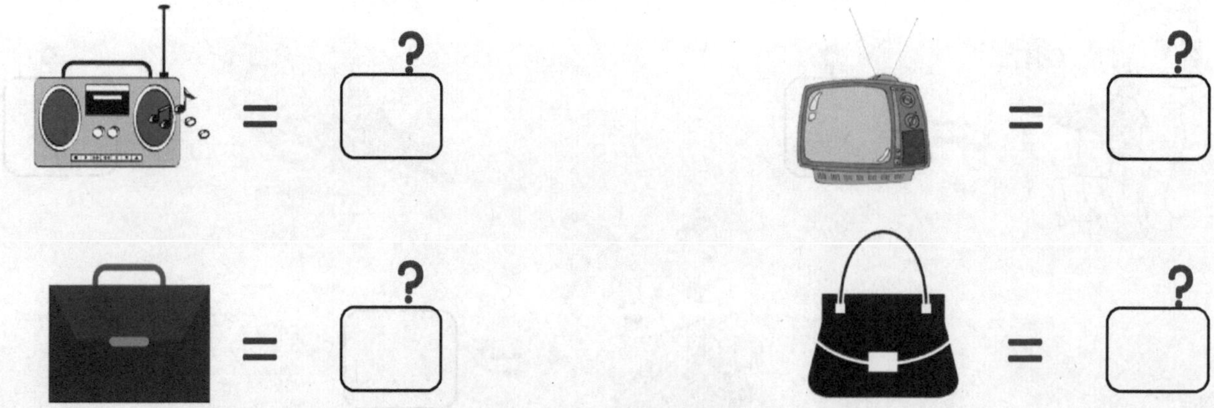

= ?

= ?

= ?

= ?

Addition : Game and Challenge

Use your math skills to find the value of each "?".

+ + = 23

+

+ + = 30

+

=

24

= ?

= ?

= ?

Addition : Game and Challenge

Use your math skills to find the value of each "?".

+ + = 22

+

+ + = 12

+

=
16

= ?

= ?

= ?

Addition : Game and Challenge

Each of the integers from 1 to 9 is to be placed in one of the circles in the figure so that the sum of the integers along each side of the figure is 20.

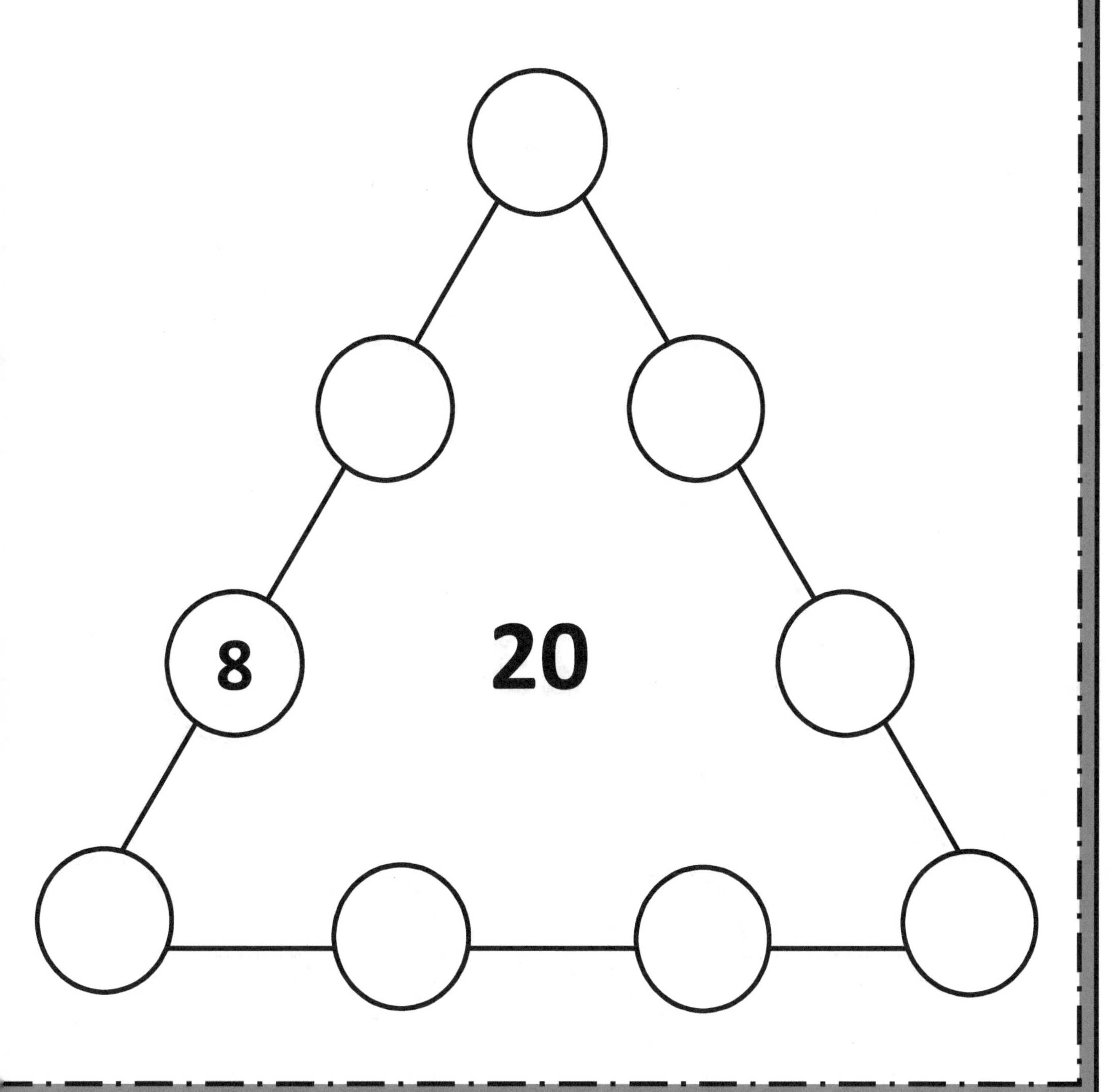

Addition : Game and Challenge

Each of the integers from 1 to 9 is to be placed in one of the circles in the figure so that the sum of the integers along each side of the figure is 23.

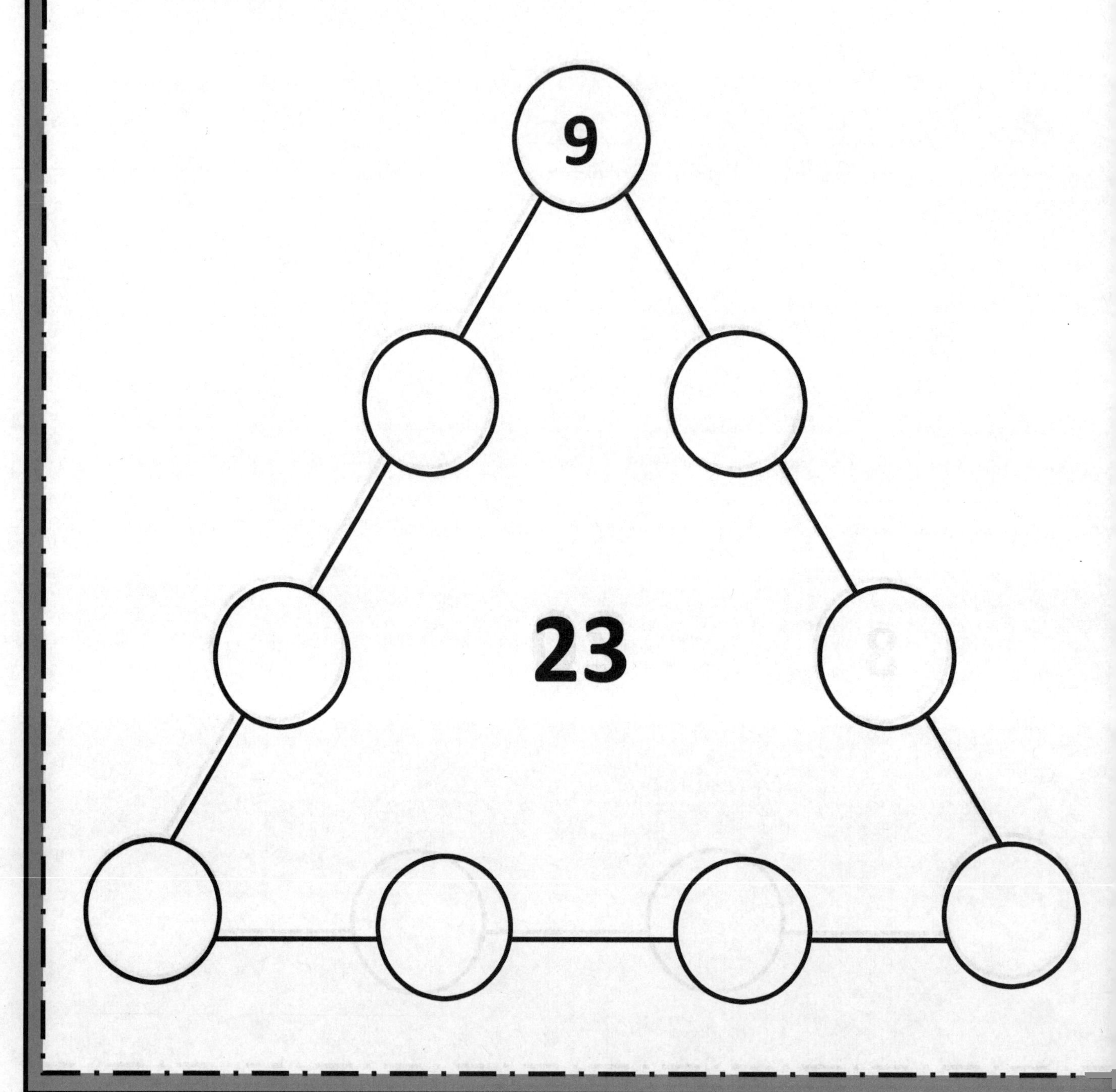

Subtracting in Columns

Subtracting 2-digit numbers (no regrouping)

Find the difference.

57 − 44	35 − 13	42 − 21	68 − 44
29 − 12	97 − 21	80 − 10	77 − 55
88 − 62	89 − 35	91 − 90	87 − 80
37 − 24	45 − 34	94 − 54	93 − 31

Subtracting in Columns

Subtracting 2-digit numbers (with regrouping)

Find the difference.

35 − 18	41 − 23	26 − 7	23 − 9
74 − 19	57 − 49	62 − 36	80 − 73
34 − 15	87 − 69	93 − 78	41 − 26
94 − 88	85 − 49	72 − 57	83 − 66

Subtracting in Columns

Subtracting 3-digit numbers (no regrouping)

Find the difference.

758 − 303	615 − 210	847 − 346	578 − 124
921 − 401	817 − 602	758 − 221	661 − 520
839 − 721	927 − 105	789 − 723	978 − 212
968 − 911	539 − 508	623 − 613	999 − 934

Subtracting in Columns

Subtracting 3-digit numbers (with regrouping)

Find the difference.

536 - 257	531 - 483	845 - 597	651 - 599
951 - 773	856 - 709	468 - 379	401 - 325
500 - 351	784 - 109	627 - 549	940 - 547
631 - 627	213 - 198	872 - 806	902 - 899

Mental Subtraction

Subtraction : Missing Number

Fill in the missing number.

1. $6 - \square = 2$

2. $7 - \square = 5$

3. $8 - \square = 3$

4. $9 - \square = 3$

5. $14 - \square = 4$

6. $17 - \square = 2$

7. $19 - \square = 5$

8. $20 - \square = 15$

9. $16 - \square = 7$

10. $18 - \square = 4$

Mental Subtraction

Subtraction : Missing Number

Fill in the missing number.

1. 40 - ☐ = 19

2. 58 - ☐ = 15

3. 36 - ☐ = 11

4. 61 - ☐ = 32

5. 73 - ☐ = 7

6. 86 - ☐ = 26

7. 90 - ☐ = 53

8. 78 - ☐ = 19

9. 84 - ☐ = 66

10. 92 - ☐ = 76

Mental Subtraction

Subtraction : Missing Number

Fill in the missing number.

1. 40 - ☐ = 20

2. 70 - ☐ = 30

3. 30 - ☐ = 20

4. 50 - ☐ = 20

5. 80 - ☐ = 30

6. 60 - ☐ = 20

7. 90 - ☐ = 10

8. 90 - ☐ = 40

9. 80 - ☐ = 70

10. 90 - ☐ = 20

Word Problems

Read and solve the problems.

1) Cathy needs at least 900 points to go to level 2 in a video game. She has only 325 points in level 1. How many more points does she need to qualify for level 2?

2) Clara bought a brand new smartphone for $679. The estimated value of the smartphoner after 5 years is $213. If she sells the smartphone after 5 years, by how much less money would she have?

3) Dora's pencil is 30 inches long. If she sharpens four inches off on Wednesday and two inches on Thursday and five inches on Friday, how long will her pencil be then?

Subtraction

Word Problems

Read and solve the problems.

1) Joe wants to practice goal kicks for soccer. He decides to have 100 kicks before going home from the stadium. He takes 51 kicks before taking a break to get a drink of water. He then takes another 27 kicks. How many more kicks does he need to make before he goes home?

2) Jacob, Michael and Jack are building towers of blocks. Jacob uses 25 blocks. Michael uses 31 blocks. Jack uses 15 blocks.
If there are 144 blocks in total, how many blocks are left?

3) Elijah found 4 small whiteboards and 7 boxes of markers for coaches to use. Oliver said they need a total of 13 whiteboards and 10 boxes of markers. How many more whiteboards and boxes of markers were needed?

Subtraction

Word Problems

Read and solve the problems.

1) James has a box of toy vehicles. There are 26 cars, 30 trucks and 42 emergency vehicles. 15 cars are blue and 9 are white. The rest are red. How many red cars are there in the box?

2) There was a small café at the corner of two streets and they served coffee and light snacks during the afternoons. There were 80 mugs for hot drinks and 60 glasses for cold drinks. One day, a waiter dropped a tray of dishes and broke 15 mugs and 6 glasses. How many glasses were left?

3) Jack is making pictures with different shapes. He cuts out some triangles, squares and circles using red, green, and blue construction paper. He cut out 29 triangles, 25 squares and 12 circles. If there are 7 red triangles and the same number of blue triangles, how many green triangles are there?

Subtraction

Word Problems

Read and solve the problems.

1) Lucas told his teacher that she made a mistake correcting his math test. She gave him a mark of 18 out of 20. It should have been 13 out of 20. How many points must the teacher now take away?

2) 51 students are going to the zoo. the students need to go in 3 groups. There are 20 students in the first group and 16 students in the second group. How many students are in the third group?

3) In the basketball arcade game, players need to get 70 points in one game to receive prizes. Partway through the game, Owen had 35 points and Samuel had 27 points. How many more points does Samuel need to receive a prize?

Add and Subtract

Add and subtract 3 single-digit numbers

Find the solution.

1. $7 + 4 - 5 = \square$

2. $6 + 5 - 8 = \square$

3. $8 + 6 - 9 = \square$

4. $7 + 7 - 6 = \square$

5. $9 + 3 - 8 = \square$

6. $9 + 4 - 9 = \square$

7. $8 + 2 - 4 = \square$

8. $8 + 0 - 2 = \square$

9. $9 + 8 - 7 = \square$

10. $8 + 5 - 4 = \square$

Add and Subtract

Add and subtract 3 single-digit numbers

Find the solution.

1. $8 + 3 - \square = 7$

2. $9 + 5 - \square = 12$

3. $10 + \square - 5 = 5$

4. $\square + 8 - 3 = 15$

5. $9 + 2 - \square = 3$

6. $\square + 5 - 1 = 17$

7. $10 + \square - 9 = 18$

8. $9 + 7 - \square = 1$

9. $10 + 8 - \square = 0$

10. $7 + 8 - \square = 5$

Add and Subtract

Word Problems

Read and solve the problems.

1) A string of lights has 19 light bulbs but 10 of them are broken. Jack only has 6 replacement bulbs. How many light bulbs are working?

2) On Thursday, 47 patients made appointments with Dr. Bloom and 17 patients did not show up. Also, 8 patients came in with no appointments. How many patients did Dr. Bloom have on Thursday?

3) Sean had fifty toy cars. If he gets twelve more cars and gives five to his little brother, how many cars will he have then?

Add and Subtract

Word Problems

Read and solve the problems.

1) Gabriel has 40 pencils. He gives 13 pencils to a friend. He buys 27 more pencils. How many pencils does Gabriel have now?

2) Lucy's mom baked 39 cookies. Brian's dad baked 26 cookies. Lucy and Brian ate 9 cookies each and then brought them to school for a party. How many cookies did they bring to school altogether?

3) 14 new lockers are delivered to the school but 6 of them come in the wrong size. The lockers with the right size are to be put on the first floor. Together with the old 3 lockers on the first floor, how many lockers are there in total on the 1st floor?

Add and Subtract

Word Problems

Read and solve the problems.

Asher, Isaac and Mila are keeping score of the game they are playing. When a player wins a game, that player gets 5 points. If a player loses a game, the player has 3 points taken away. If it is a tie, every player gets 2 points. Each of them has 20 points to start with. Asher wins the first game.

1) Isaac wins the second game. How many points does Isaac has after the second game? (remember to count the points Isaac gets for the first game!)

2) The third game is a tie. How many points does Mila have after the third game?

Add and Subtract

Word Problems

Read and solve the problems.

A bus leaves the terminal every morning at 7 o'clock.
There are 30 seats on the left side of the bus and 26 seats on the right side. When the bus left the terminal this morning, 8 passengers sat on the bus. At the first stop, 17 passengers got on the bus.

1) At the second stop, 13 passengers get on the bus and 10 got off. How many passengers were there on the bus?

2) There were 5 less passengers getting on at the third stop than the second stop. 4 passengers got off at the third stop. How many passengers were there on the bus after the 3rd stop?

Add & Subtract : Game and Challenge

Use your math skills to find the value of each "?".

\- + = 12

\+ = 8

= ? = ?

\- = 10

\+ - = 0

= ? = ?

Add & Subtract : Game and Challenge

Use your math skills to find the value of each "?".

+ + = 27

+ = 14

- = 5

= ?

= ?

= ?

Add & Subtract : Game and Challenge

Use your math skills to find the value of each "?".

triangle	+	2	=	moon
moon	-	9	=	star
18	-	star	=	heart
heart	+	4	=	15

moon = ?

triangle = ?

heart = ?

star = ?

Add & Subtract : Game and Challenge

Use your math skills to find the value of each "?".

★ + ★ - ▬ = 34

★ - ♥ = 10

☾ = 32-21+6

▬ + ☾ = 23

☾ = ?

★ = ?

▬ = ?

♥ = ?

Add & Subtract : Game and Challenge

Use your math skills to find the value of each "?".

\- 5 =

\+ 6 =

\+ 3 = 7

= ?

= ?

= ?

= ?

Add & Subtract : Game and Challenge

Use your math skills to find the value of each "?".

☆ + ♡ + ☆ = ▭
25 - ▭ = ☾
☾ + ☾ = 10
☆ + ☆ = ☾ + 9
▭ - ♡ = ?

☾ = ?

☆ = ?

▭ = ?

♡ = ?

Add & Subtract : Game and Challenge

Use your math skills to find the value of each "?".

+  +  +  = 31

19 =  +

+  - 3 = 13

30 -  = 18

= ?

= ?

= ?

= ?

Add & Subtract : Game and Challenge

Use your math skills to find the value of each "?".

\- 3 + = 19

5 = -

\- 7 =

\+ - + = ?

= ?

= ?

= ?

= ?

Add & Subtract : Game and Challenge

Use your math skills to find the value of each "?".

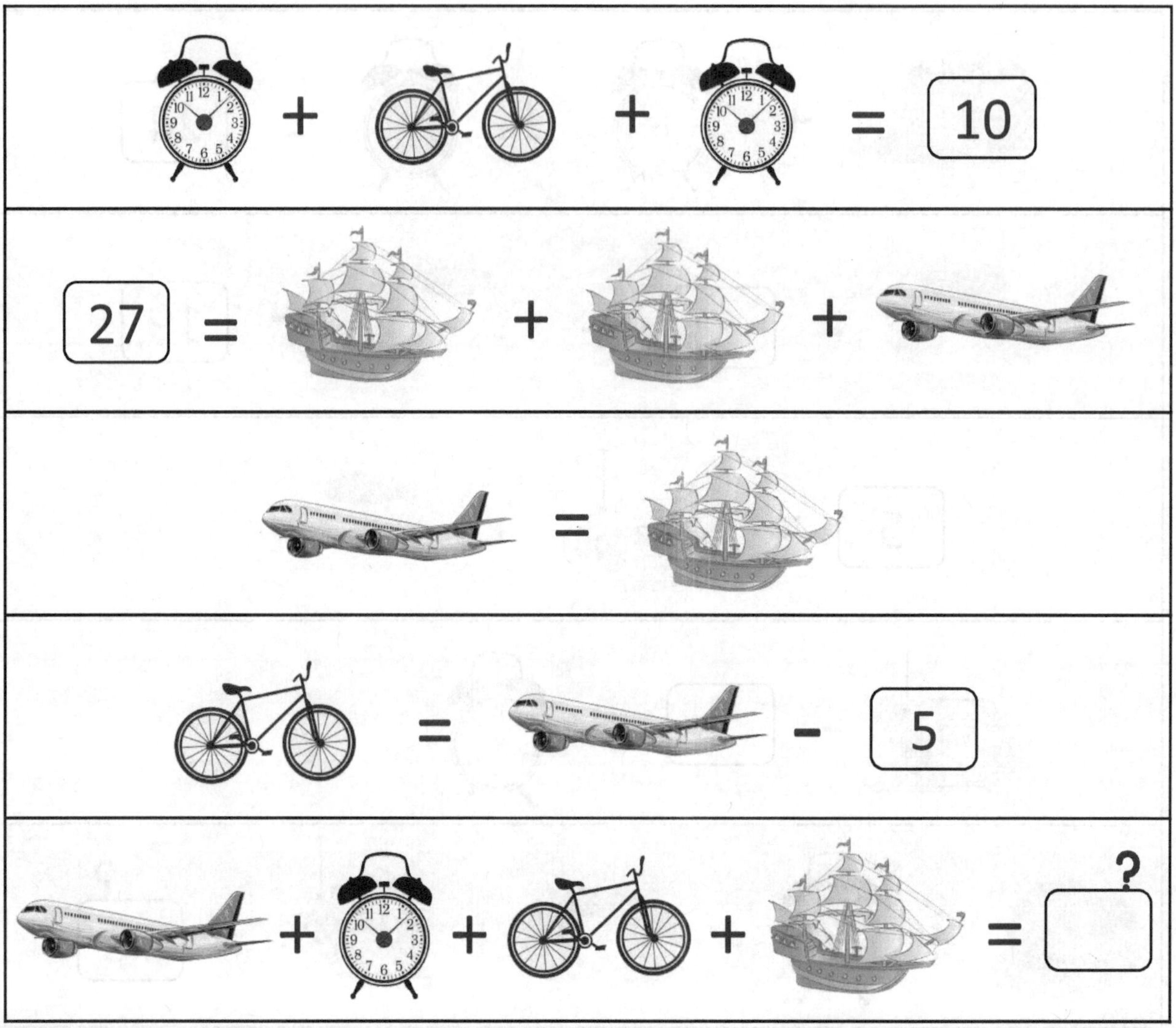

+ + = 10

27 = + +

=

= - 5

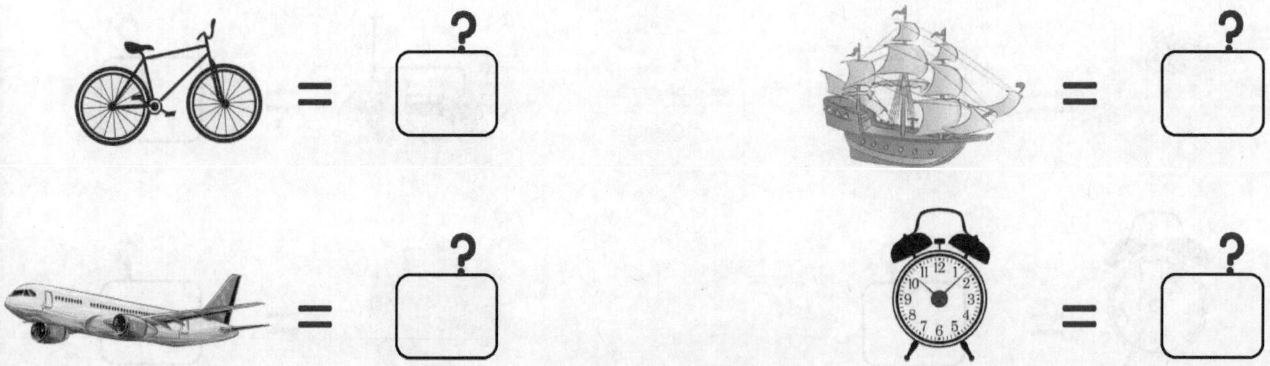

+ + + = ?

= ?

= ?

= ?

= ?

Multiplication

Meaning of Multiplication

Multiplying means repeated addition of a number. (The number must all be the same before we can use it to multiply).

Example:

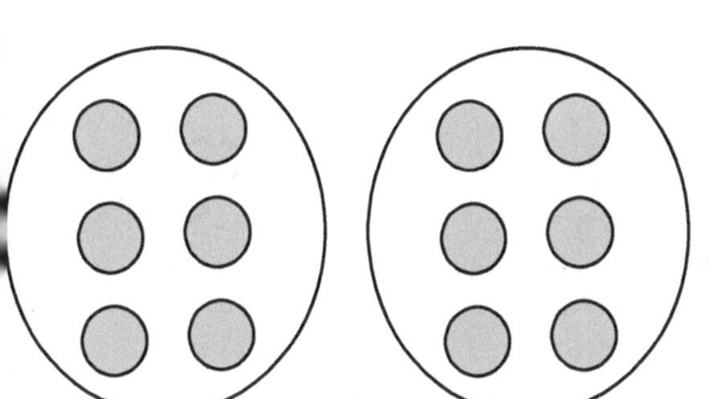

2 groups of 6

Or 6 + 6

Or 2 x 6

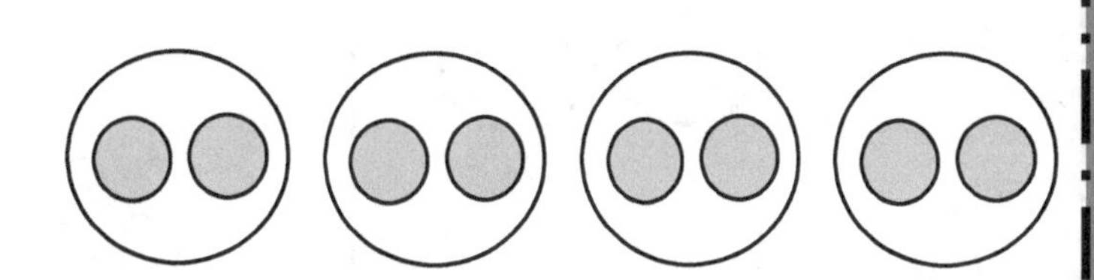

4 groups of 2

Or 2 + 2 + 2 +2

Or 4 x 2

Here is another example :

3 x 7 = 7 + 7 + 7 = 21

5 x 6 = 6 + 6 + 6 + 6 + 6 = 30

Multiplication

Meaning of Multiplication

Circle the group of objects that match the equation:

$2 \times 2 = 4$

$4 \times 3 = 12$

Multiplication

Meaning of Multiplication

Circle the group of objects that match the equation:

$3 \times 4 = 12$

$6 \times 2 = 6$

Multiplication

Meaning of Multiplication

Fill in the blanks.

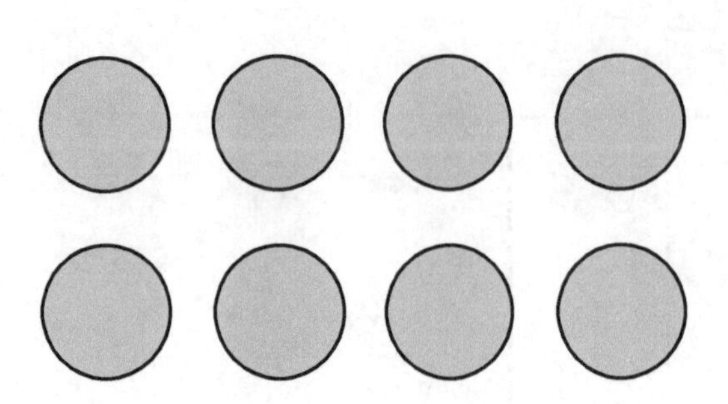

2 rows of **4**

2 x 4 = 8

___ rows of ___

___ x ___ = ___

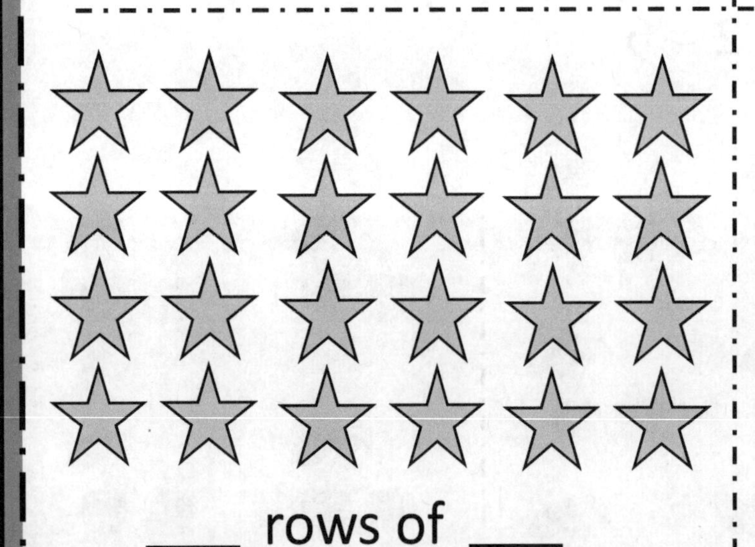

___ rows of ___

___ x ___ = ___

___ rows of ___

___ x ___ = ___

Multiplication

Meaning of Multiplication

Fill in the blanks.

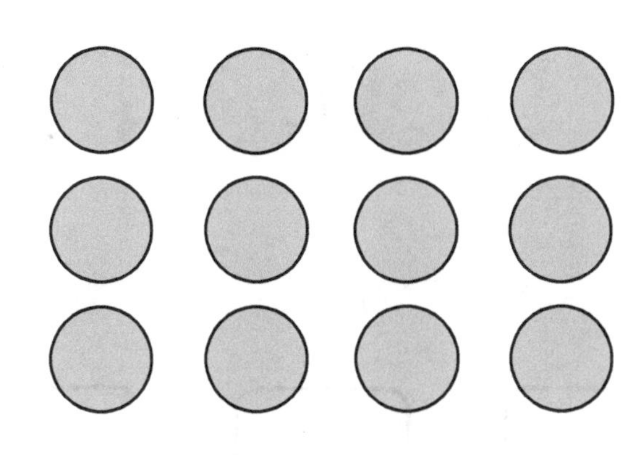

___ rows of ___

___ x ___ = ___

___ rows of ___

___ x ___ = ___

___ rows of ___

___ x ___ = ___

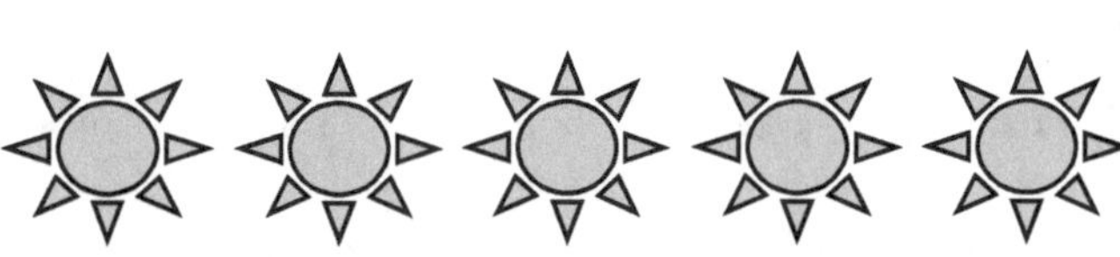

___ rows of ___

___ x ___ = ___

Multiplication

Meaning of Multiplication

Complete the equation to describe the array.

3 × 2 = 6

1 × ☐ = ☐

2 × ☐ = ☐

☐ × 4 = ☐

Multiplication

Meaning of Multiplication

Complete the equation to describe the array.

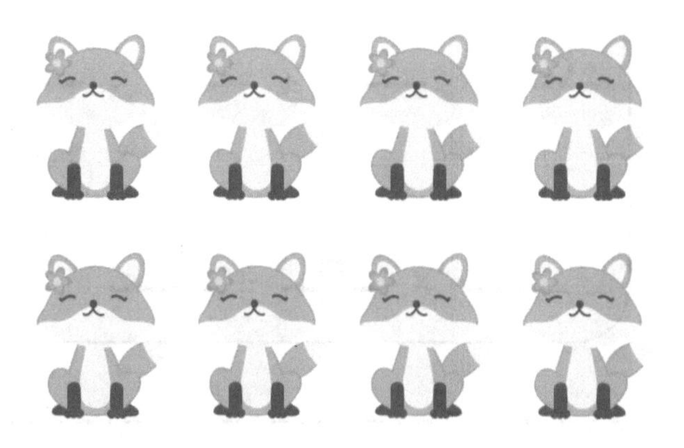

□ × 2 = □

1 × □ = □

3 × □ = □

□ × 4 = □

Multiplication

Meaning of Multiplication

Fill in the blanks.

6 + ___ + ___ = ___

3 x 6 = ___

8 + ___ = ___

___ x ___ = ___

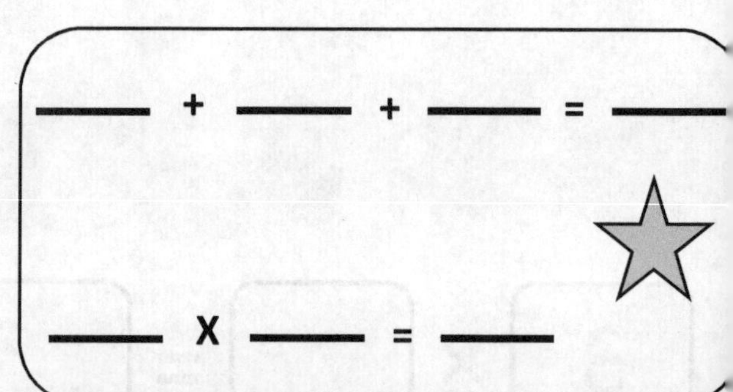

___ + ___ + ___ = ___

___ x ___ = ___

Multiplication Tables

Multiplication tables of 2 and 3

Find the product.

1. $2 \times 3 = \square$

2. $2 \times 9 = \square$

3. $5 \times 2 = \square$

4. $6 \times 2 = \square$

5. $7 \times 2 = \square$

6. $4 \times 3 = \square$

7. $3 \times 3 = \square$

8. $5 \times 3 = \square$

9. $7 \times 3 = \square$

10. $3 \times 6 = \square$

Multiplication Tables

Multiplication tables of 4 and 5

Find the product.

1. $4 \times 2 = \square$
2. $4 \times 5 = \square$
3. $5 \times 6 = \square$
4. $5 \times 5 = \square$
5. $3 \times 4 = \square$
6. $4 \times 4 = \square$
7. $8 \times 5 = \square$
8. $7 \times 4 = \square$
9. $6 \times 4 = \square$
10. $5 \times 9 = \square$

Multiplication Tables

Multiplication tables of 6 and 7

Find the product.

1. $6 \times 2 = \square$

2. $3 \times 7 = \square$

3. $7 \times 5 = \square$

4. $6 \times 6 = \square$

5. $8 \times 6 = \square$

6. $7 \times 6 = \square$

7. $7 \times 7 = \square$

8. $7 \times 8 = \square$

9. $9 \times 6 = \square$

10. $7 \times 9 = \square$

Multiplication Tables

Multiplication tables of 8 and 9

Find the product.

1. 8 x 3 =

2. 9 x 5 =

3. 5 x 8 =

4. 9 x 1 =

5. 4 x 9 =

6. 8 x 8 =

7. 9 x 6 =

8. 9 x 8 =

9. 9 x 9 =

10. 8 x 4 =

Multiplication

Word Problems

The class is doing a math activity. There are 6 groups of 3 students.

1. How many students are there in the class?

2. Each group should have 2 pairs of pens. How many pairs of pens are needed in total?

3. Each student should get 5 worksheets and 4 sheets of construction paper. How many sheets of construction paper would each group of students have?

4. Each student needs to answer 7 questions on each worksheet. How many questions does each student need to answer?

5. Student get 4 stickers on their reward chart for each correct answer. If a student gets 8 correct answers, how many stickers will he get?

Multiplication

Word Problems

There are 6 houses on a small street. Each house has a family with 4 children, 2 cats and a dog.

1. How many children are there living on this street?

2. How many pets are there living on this street?

3. Outside each house, there is a mailbox. There are 4 letters in each of the six mailboxes. How many letters are there in total?

4. Last Saturday, two of the houses were having a party. Each family invited 9 guests to their house. How many guests were there in total?

5. Last Sunday, 4 of the houses had some cars parked outside. If there were 4 cars outside each house. How many cars were there in total?

Multiplication

Word Problems

Ms. Hudson is getting her cooking class ready. She has 4 students in this class and they are going to make double chocolate muffins.

1. Ms. Hudson printed two sheets of recipes for each student. How many sheets did she print in total?

2. Each student will be making 5 muffins. Together with the 5 muffins that Ms. Hudson will be making, how many muffins will be made in total?

3. Ms. Hudson measured 4 cups of flour for each student. How many cups of flour does she measured for all the students?

4. She takes out 3 boxes of eggs from the fridge. If there are 6 eggs in each box, how many eggs are there?

5. Each bar of chocolate can be broken into 10 cubes. How many cubes are there if there are 2 bars of chocolate?

Multiplication : Game and Challenge

Use your math skills to find the value of each "?".

+ − = 4

× = 4

= ?

= ?

× = 6

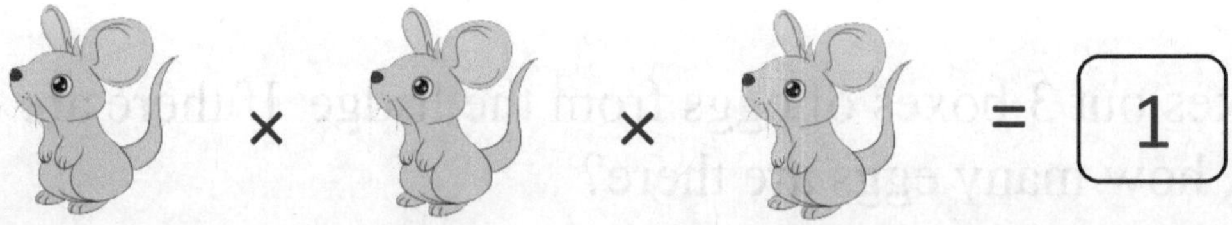

× × = 1

= ?

= ?

Multiplication : Game and Challenge

Use your math skills to find the value of each "?".

× × = 36

× = 9

= ?

= ?

× = 16

× - = 10

= ?

= ?

Multiplication : Game and Challenge

Fill the missing numbers to complet the product.

2	×	◯	=	6
×		×		×
◯	×	1	=	◯
=		=		=
◯	×	◯	=	18

◯	×	◯	=	4
×		×		×
◯	×	3	=	◯
=		=		=
8	×	◯	=	24

Multiplication : Game and Challenge

Fill the missing numbers to complet the product.

	×		=	8
×		×		×
	×	2	=	
=		=		=
4	×		=	16

	×	2	=	6
×		×		×
5	×		=	
=		=		=
	×		=	30

Place Value

Example: $725 = 7 \times 100 + 2 \times 10 + 5 \times 1$

Write the number in expanded form.

226 = ______________________ 455 = ______________________

312 = ______________________ 563 = ______________________

104 = ______________________ 300 = ______________________

815 = ______________________ 707 = ______________________

678 = ______________________ 640 = ______________________

555 = ______________________ 901 = ______________________

760 = ______________________ 919 = ______________________

804 = ______________________ 999 = ______________________

781 = ______________________ 602 = ______________________

Place Value

Example: 725 = 7 X 100 + 2 X 10 + 5 X 1

Write each number in normal form.

1. ______ = **4 x 100 + 2 x 10 + 6 x 1**
2. ______ = **2 x 100 + 5 x 10 + 7 x 1**
3. ______ = **1 x 100 + 6 x 10 + 5 x 1**
4. ______ = **5 x 100 + 6 x 10 + 2 x 1**
5. ______ = **3 x 100 + 7 x 10 + 1 x 1**
6. ______ = **4 x 100 + 4 x 10 + 4 x 1**
7. ______ = **6 x 100 + 8 x 10 + 3 x 1**
8. ______ = **7 x 100 + 7 x 10 + 7 x 1**
9. ______ = **8 x 100 + 5 x 10 + 8 x 1**
10. ______ = **8 x 100 + 9 x 10 + 0 x 1**
11. ______ = **9 x 100 + 3 x 10 + 6 x 1**
12. ______ = **4 x 100 + 0 x 10 + 8 x 1**
13. ______ = **8 x 100 + 6 x 10 + 1 x 1**
14. ______ = **7 x 100 + 9 x 10 + 9 x 1**
15. ______ = **9 x 100 + 0 x 10 + 5 x 1**
16. ______ = **8 x 100 + 8 x 10 + 8 x 1**
17. ______ = **7 x 100 + 0 x 10 + 7 x 1**
18. ______ = **6 x 100 + 6 x 10 + 0 x 1**

Place Value

Hundreds, Tens & Ones

Determine the value of the underlined digit.

325 → 3 hundreds

572 → 2 ones

203 →

727 →

900 →

23 →

670 →

40 →

641 → 4 tens

474 →

616 →

888 →

50 →

1 →

8 →

780 →

Rounding

Round 2-digit numbers to the nearest 10

Round to the nearest ten.

36	→	______	52	→	______
47	→	______	15	→	______
8	→	______	79	→	______
14	→	______	90	→	______
43	→	______	57	→	______
88	→	______	65	→	______
34	→	______	21	→	______
70	→	______	81	→	______

Rounding

Round 3-digit numbers to the nearest 10

Round to the nearest ten.

316	→ ______	422	→ ______
777	→ ______	661	→ ______
885	→ ______	301	→ ______
705	→ ______	900	→ ______
802	→ ______	644	→ ______
777	→ ______	412	→ ______
999	→ ______	100	→ ______
707	→ ______	880	→ ______

Rounding

Round 3-digit numbers to the nearest 100

Round to the nearest hundred.

125 → __________

371 → __________

610 → __________

742 → __________

155 → __________

850 → __________

961 → __________

700 → __________

606 → __________

489 → __________

720 → __________

950 → __________

508 → __________

849 → __________

999 → __________

473 → __________

Comparing Numbers

Comparing numbers up to 1000

Write the correct sign < , > or = .

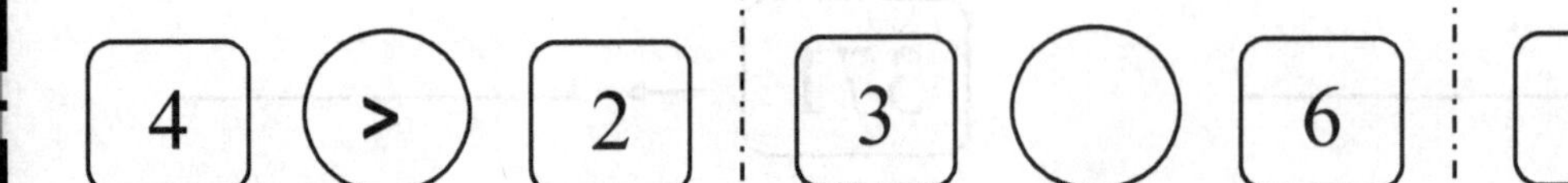

4	>	2	3		6	7		8
14		15	19		20	6		31
66		46	71		38	87		90
90		100	327		80	606		660
654		7	91		441	748		800
16		330	900		899	200		49
160		149	451		560	379		402

Comparing Numbers

Ordering numbers up to 100

Arrange these numbers in order, from least to greatest.

a. 6, 18, 4, 27	____ < ____ < ____ < ____
b. 15, 7, 21, 30	____ < ____ < ____ < ____
c. 35, 45, 71, 9	____ < ____ < ____ < ____
d. 62, 57, 33, 90	____ < ____ < ____ < ____
e. 22, 39, 51, 87	____ < ____ < ____ < ____
f. 70, 40, 90, 50	____ < ____ < ____ < ____
g. 19, 39, 90, 100	____ < ____ < ____ < ____
h. 86, 100, 22, 73	____ < ____ < ____ < ____

Comparing Numbers

Ordering numbers up to 1,000

Arrange these numbers in order, from least to greatest.

a. 348, 11, 102, 96	____ < ____ < ____ < ____
b. 107, 303, 444, 500	____ < ____ < ____ < ____
c. 610, 120, 25, 320	____ < ____ < ____ < ____
d. 130, 745, 881, 666	____ < ____ < ____ < ____
e. 500, 900, 200, 1000	____ < ____ < ____ < ____
f. 120, 657, 107,7	____ < ____ < ____ < ____
g. 925, 911, 1000, 852	____ < ____ < ____ < ____
h. 1000, 654, 360, 479	____ < ____ < ____ < ____

Time

Telling Time

Draw the time shown on each clock.

1.

3:00

2.

2:35

3.

1:30

4.

7:25

5.

10:30

6.

8:15

7.
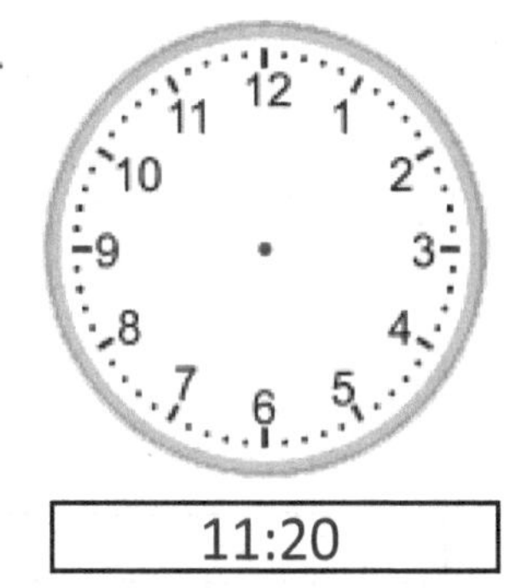
11:20

8.

2:30

9.

9:00

Time

Telling Time

Draw the time shown on each clock.

1.

2.
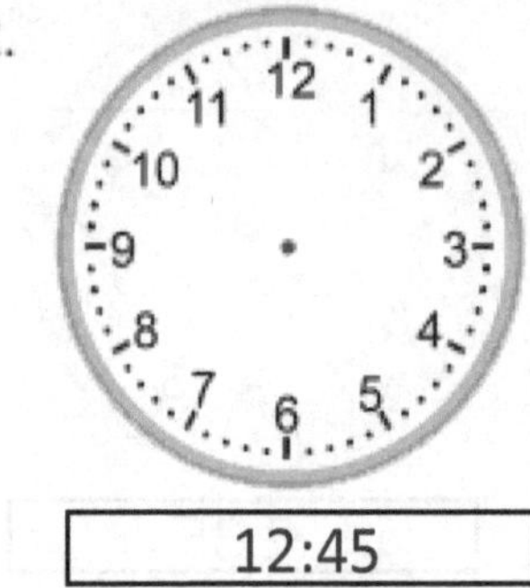

3.
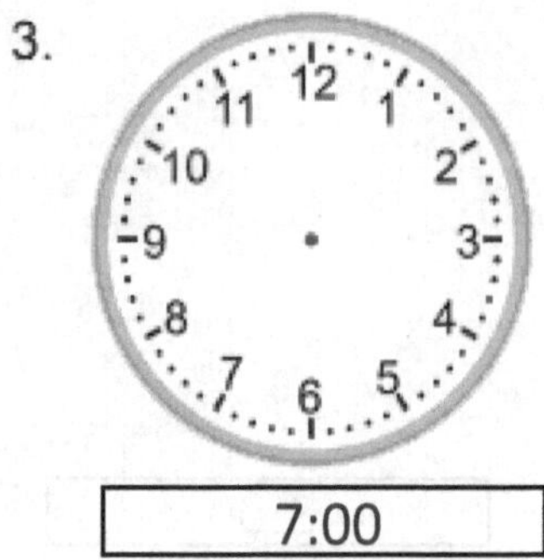

4.
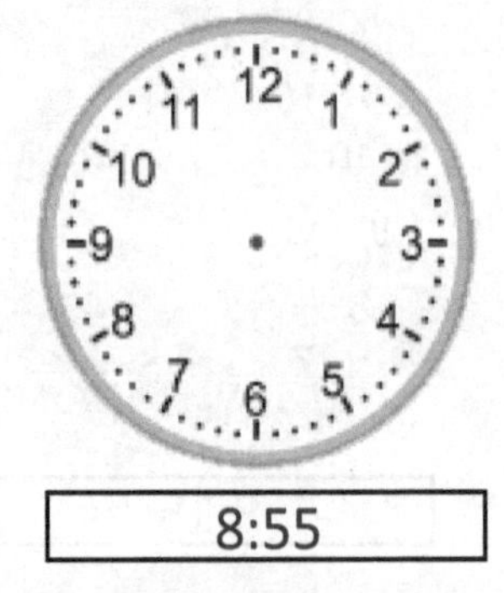

5.
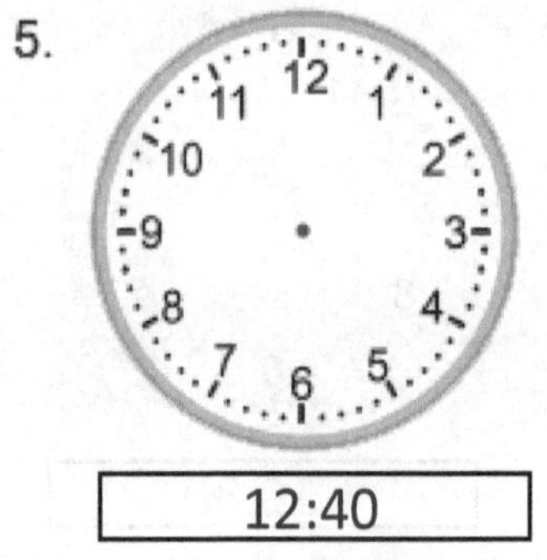

6.
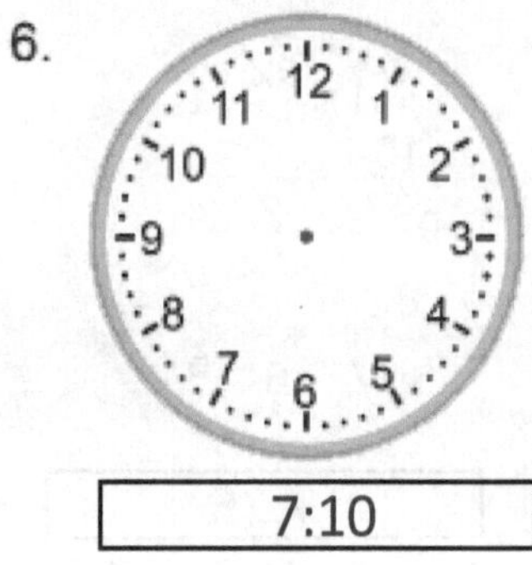

7.
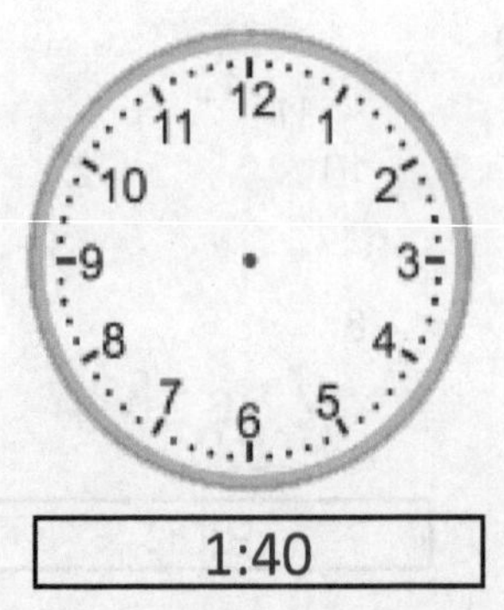

8.
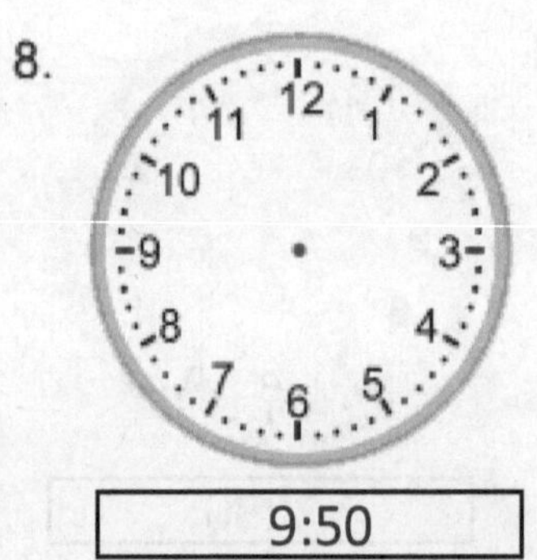

9.
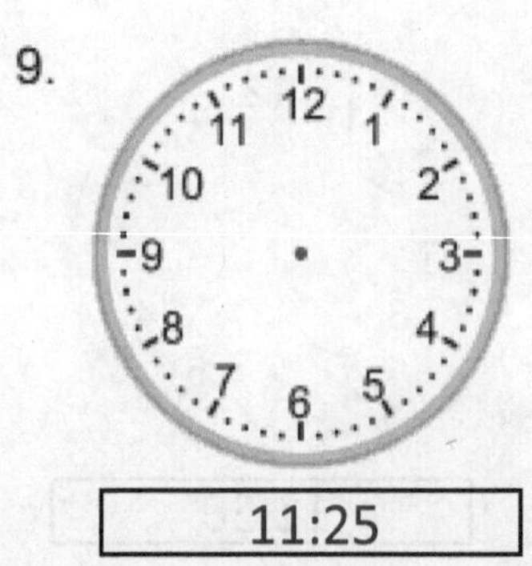

Time

Units of Time

Circle the best estimate of the time needed for each activity.

1) Filling up a water bottle.

30 seconds   30 minutes   30 hours

2) Having a summer holiday from school.

2 years   2 months   2 days

3) sleeping at night.

Weeks   Minutes   Hours

4) Making a cup of coffee.

2 seconds   2 hours   2 minutes

5) Counting from 1 to 20.

20 minutes   20 seconds   20 weeks

Units of Time

Circle the best estimate of the time needed for each activity.

1) Preparing dinner.

Months Years Minutes

2) Taking a picture.

Years Seconds Months

3) Brushing your teeth.

1 week 1 minute 1 year

4) Building a new bridge.

3 days 3 months 3 minutes

5) Reading.

Seconds Minutes Weeks

Time

Elapsed Time

Draw the clock hands to show the time it was or will be.

1)  What time will it be in 1 hour 0 minutes?	2)  What time will it be in 3 hours 0 minutes?
3)  What time was it 4 hours 0 minutes ago?	4)  What time was it 5 hours 0 minutes ago?
5)  What time will it be in 4 hours 30 minutes?	6)  What time was it 2 hours 30 minutes ago?

Time

Elapsed Time

Draw the clock hands to show the time it was or will be.

<table>
<tr>
<td>1)
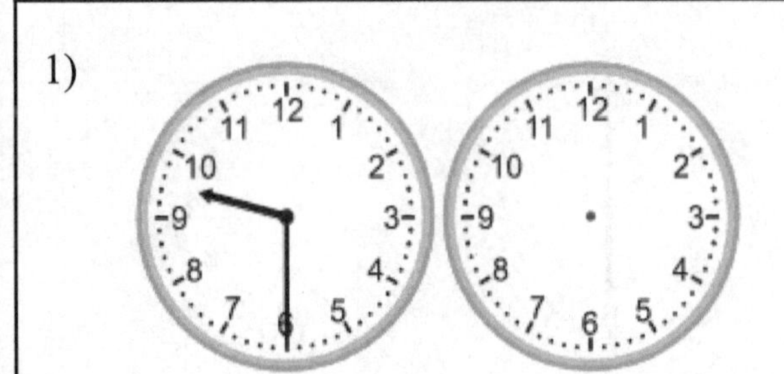
What time will it be in 3 hours 25 minutes?</td>
<td>2)

What time will it be in 4 hours 40 minutes?</td>
</tr>
<tr>
<td>3)

What time was it 2 hours 45 minutes ago?</td>
<td>4)
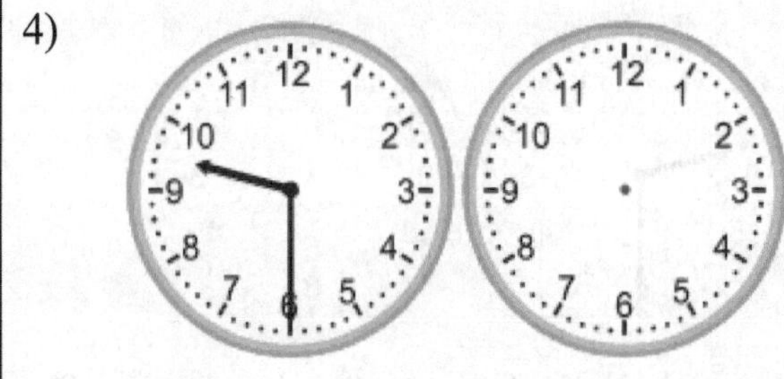
What time was it 4 hours 50 minutes ago?</td>
</tr>
<tr>
<td>5)

What time will it be in 5 hours 15 minutes?</td>
<td>6)

What time will it be in 6 hours 50 minutes?</td>
</tr>
</table>

Time

Word Problems

Read and answer each question.

The aquarium is the favourite place of all the kids in town!

1. The aquarium opens at 10 o'clock in the morning and closes at 8:30 in the evening. How many hours is it open?

2. The dolphins do performances every four hours. They did one show at 11:00. At what time is the next dolphin show?

3. Kids can pat starfish at the starfish exhibit everyday at 11 o'clock in the morning. The patting session will last for one and a half hours. When is the patting session done?

4. A special tour was scheduled to start at 5 o'clock in the afternoon but started a half an hour late. If the tour ended at half past seven in the evening, how long did the tour last?

5. The ticketing office stops selling tickets half an hour before the closing time of the aquarium. At what time does the ticketing office close?

Time

Word Problems

Read and answer each question.

Ashley was babysitting Mrs. Bloom's twins last night.

1. Ashley started babysitting at 5 o'clock in the afternoon. Mrs. Bloom was supposed to come home at 8:30. How long did Ashley expect to babysit for?

2. Mrs. Bloom called at a 7:15 in the evening and said she could not be back until two hours later. At what time would she be back?

3. How much later was Mrs. Bloom than the original schedule?

4. The twins did not want to go to bed! Ashley finally got them into bed at 9:10, which is 40 minutes later than their usual bed time. When is their usual bedtime?

5. It took Ashley 25 minutes to get home from Mrs. Bloom's house. She left Mrs. Bloom's house at half past nine. At what time did Ashley get home?

Measurement

Estimate and measure length

Estimate the height of each picture in inches and centimeters.

Pine tree

Height: ______ inches

Height: ______ cm

Maple tree

Height: ______ inches

Height: ______ cm

Measure the height of the pictures using a ruler.

Pine tree

Height: ______ inches

Height: ______ cm

Maple tree

Height: ______ inches

Height: ______ cm

Measurement

Estimate and measure length

Estimate the height of each picture in inches and centimeters.

Boy

Height: ______ inches

Height: ______ cm

Man

Height: ______ inches

Height: ______ cm

Measure the height of the pictures using a ruler.

Boy

Height: ______ inches

Height: ______ cm

Man

Height: ______ inches

Height: ______ cm

Measurement

Estimate and measure length

Estimate the height of each picture in inches and centimeters.

Horse

Height: ______ inches

Height: ______ cm

Height: ______ inches

Height: ______ cm

Measure the height of the pictures using a ruler.

Horse

Height: ______ inches

Height: ______ cm

Dog

Height: ______ inches

Height: ______ cm

Measurement

Units of length : inches, feet, yards & miles

Circle the proper unit for each of the following.

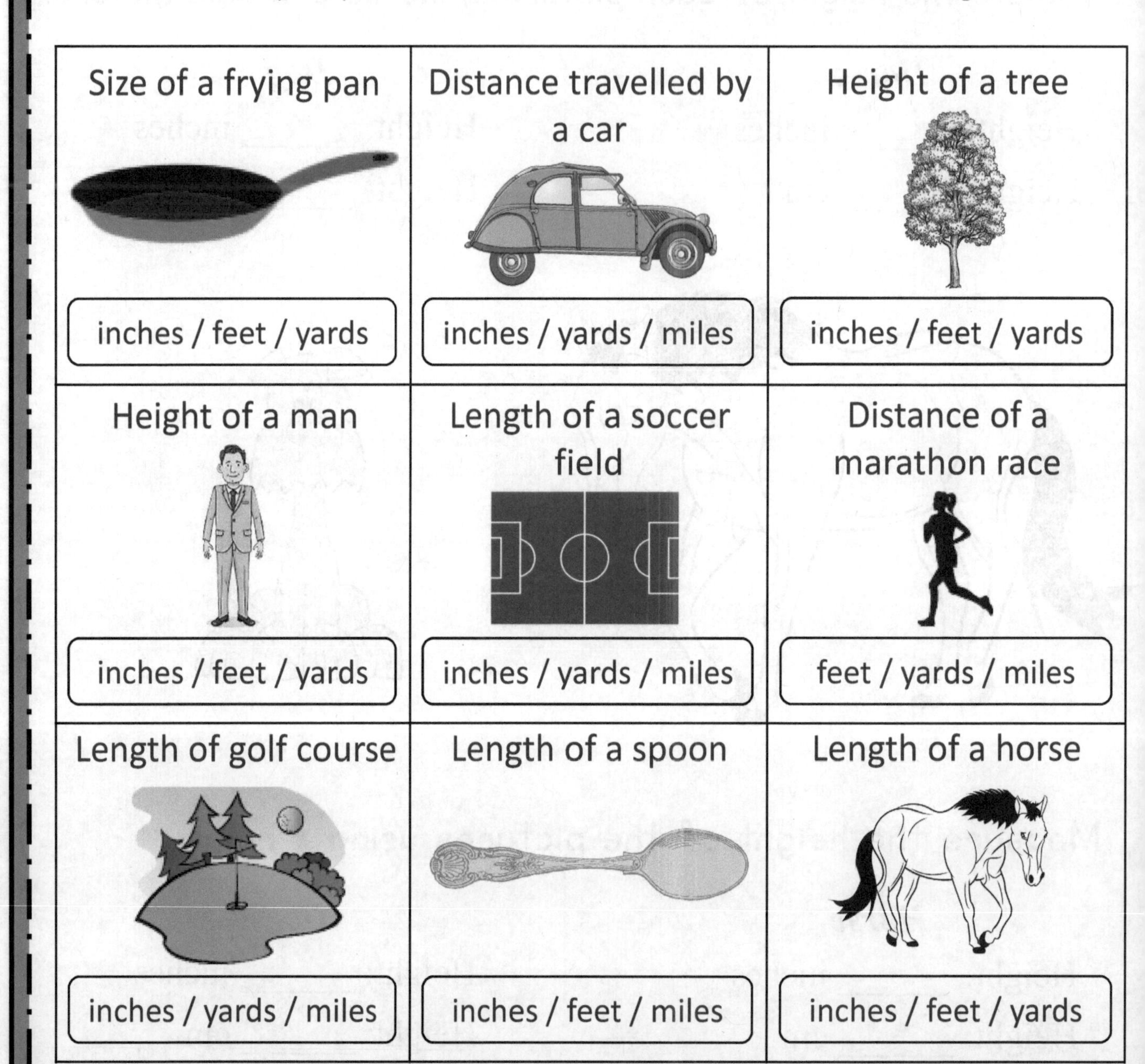

Size of a frying pan	Distance travelled by a car	Height of a tree
inches / feet / yards	inches / yards / miles	inches / feet / yards
Height of a man	**Length of a soccer field**	**Distance of a marathon race**
inches / feet / yards	inches / yards / miles	feet / yards / miles
Length of golf course	**Length of a spoon**	**Length of a horse**
inches / yards / miles	inches / feet / miles	inches / feet / yards

Measurement

Units of length : centimeters, meters and kilometers

Circle the proper unit for each of the following.

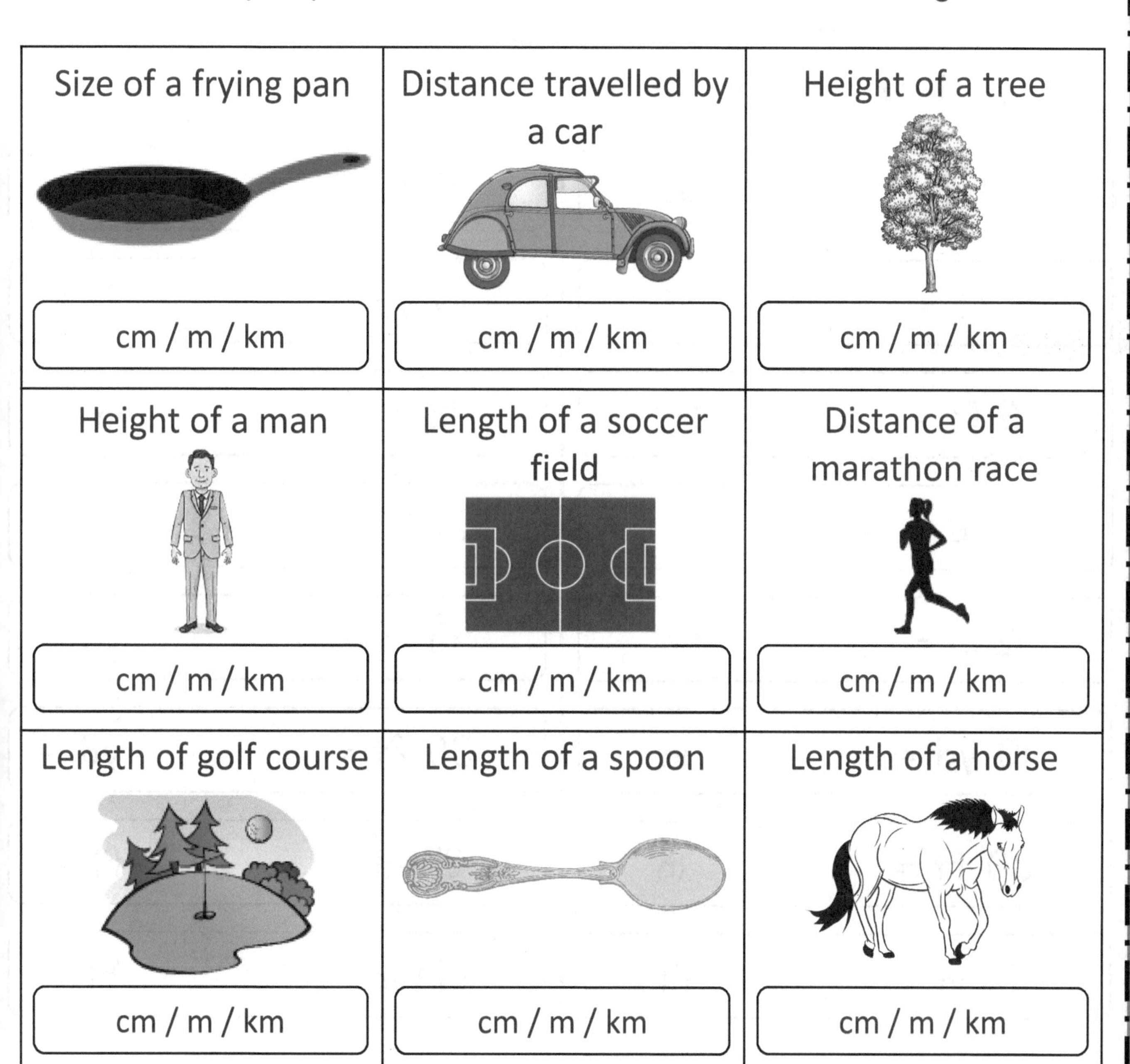

Size of a frying pan	Distance travelled by a car	Height of a tree
cm / m / km	cm / m / km	cm / m / km
Height of a man	**Length of a soccer field**	**Distance of a marathon race**
cm / m / km	cm / m / km	cm / m / km
Length of golf course	**Length of a spoon**	**Length of a horse**
cm / m / km	cm / m / km	cm / m / km

Measurement

Convert between yards, feet and inches

Note: 1 yard (yd) = 3 feet (ft); 1 foot = 12 inches (in)

Convert to the units shown.

5 yd = ____ in	12 ft = ____ in
9 ft = ____ yd	10 yd = ____ ft
6 ft = ____ in	18 ft = ____ in
8 yd = ____ ft	27 ft = ____ yd
12 in = ____ ft	45 ft = ____ yd
10 yd = ____ in	36 ft = ____ yd
30 ft = ____ in	21 yd = ____ ft
36 in = ____ ft	144 in = ____ ft

Measurement

Convert between meters, centimeters and millimeters

Note : 1 m = 100 cm = 1,000 mm

Convert to the units shown.

7 m = ____ cm

25 m = ____ cm

6 m = ____ mm

5 m = ____ mm

500 mm = ____ cm

600 cm = ____ m

3000 cm = ____ m

4000 mm = ____ m

3 cm = ____ mm

14 cm = ____ mm

30 cm = ____ mm

22 m = ____ mm

60 mm = ____ cm

1000 cm = ____ m

200 mm = ____ cm

1000 mm = ____ m

Units of weight - ounces, pounds & tons

Match the proper unit of measurement with the objects.

OUNCES POUNDS TONS

Measurement

Units of weight - grams & kilograms

Match the proper unit of measurement with the objects.

GRAMS

KILOGRAMS

Measurement

Convert between ounces and pounds

Note: 1 pound (lb) = 16 ounces (oz)

Convert the given measures to new units.

3 lb = oz	2 lb = oz
10 lb = oz	20 lb = oz
8 lb = oz	30 lb = oz
7 lb = oz	40 lb = oz
16 oz = lb	48 oz = lb
32 oz = lb	64 oz = lb
96 oz = lb	80 oz = lb
112 oz = lb	0 oz = lb

Measurement

Convert between kilograms and grams

Note: 1 kilogram (kg) = 1,000 grams (gm)

Convert the given measures to new units.

3 kg = g	2 kg = g
4 kg = g	6 kg = g
5 kg = g	8 kg = g
7 kg = g	9 kg = g
1000 g = kg	3000 g = kg
2000 g = kg	5000 g = kg
4000 g = kg	7000 g = kg
9000 g = kg	0 g = kg

Measurement : Game and Challenge

Use your math skills to find the value of each "?".

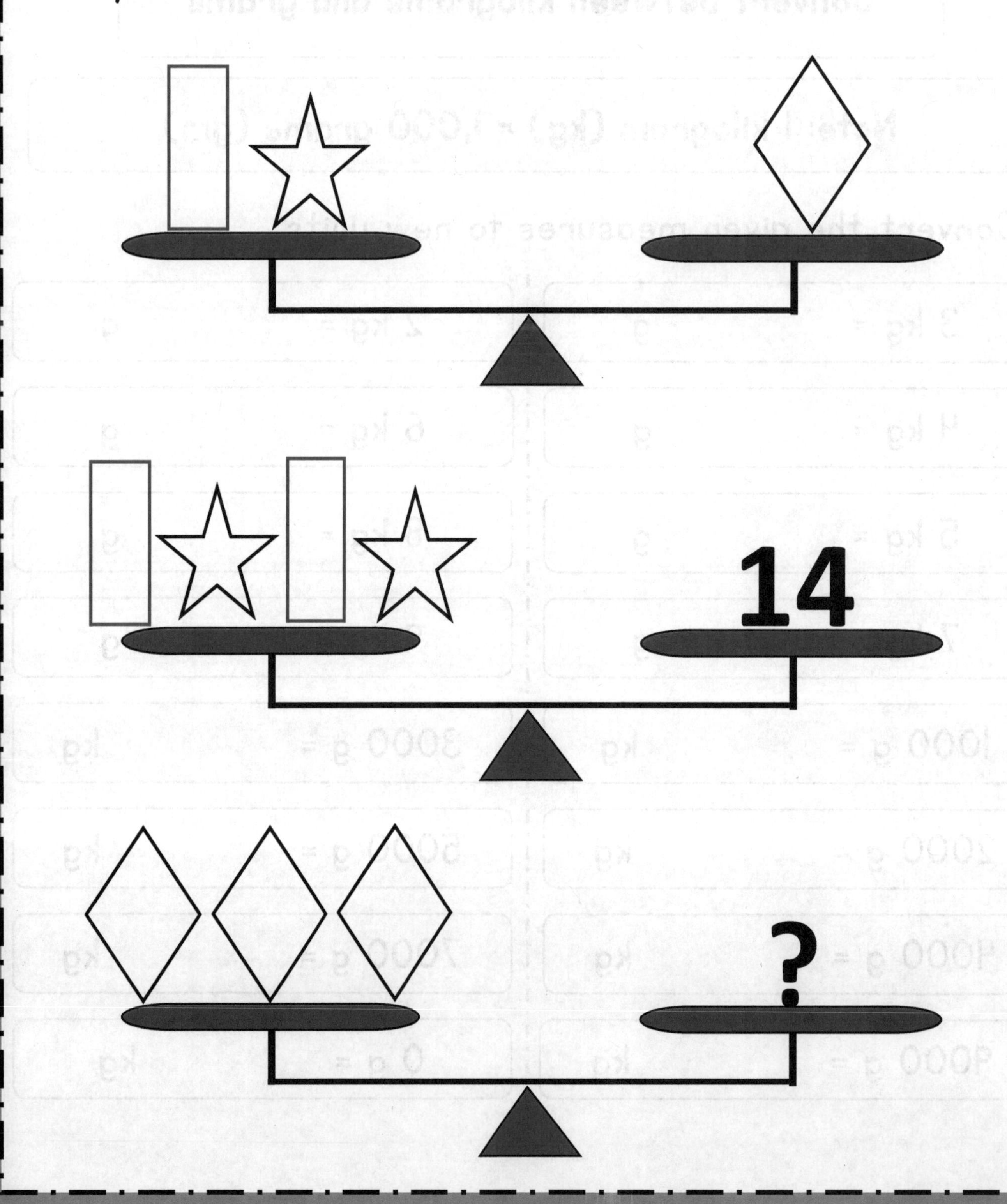

Measurement : Game and Challenge

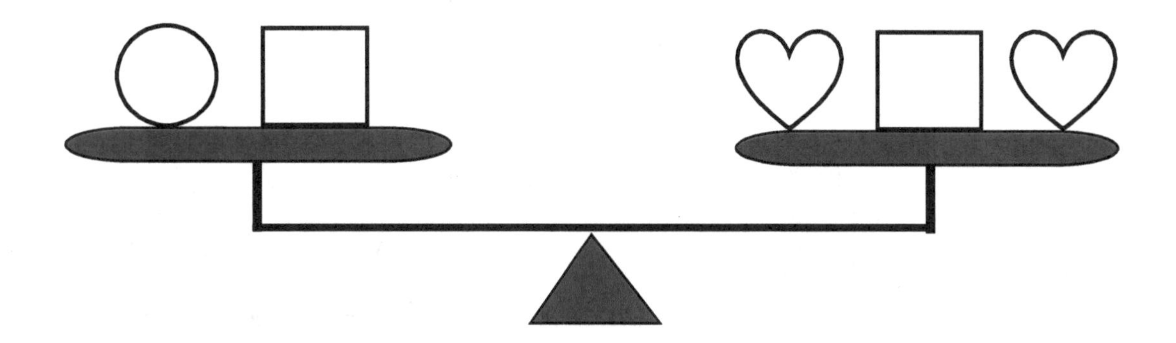

Circle the three answers that will always be true

a. ○ ♡ = □

b. ♡ ♡ = □

c. ○ = ♡ ♡

d. ♡ □ ♡ □ = □ ○ □

e. ♡ = ◡

f. □ □ = ◡ ○

Measurement : Game and Challenge

Use your math skills to find the value of each "?".

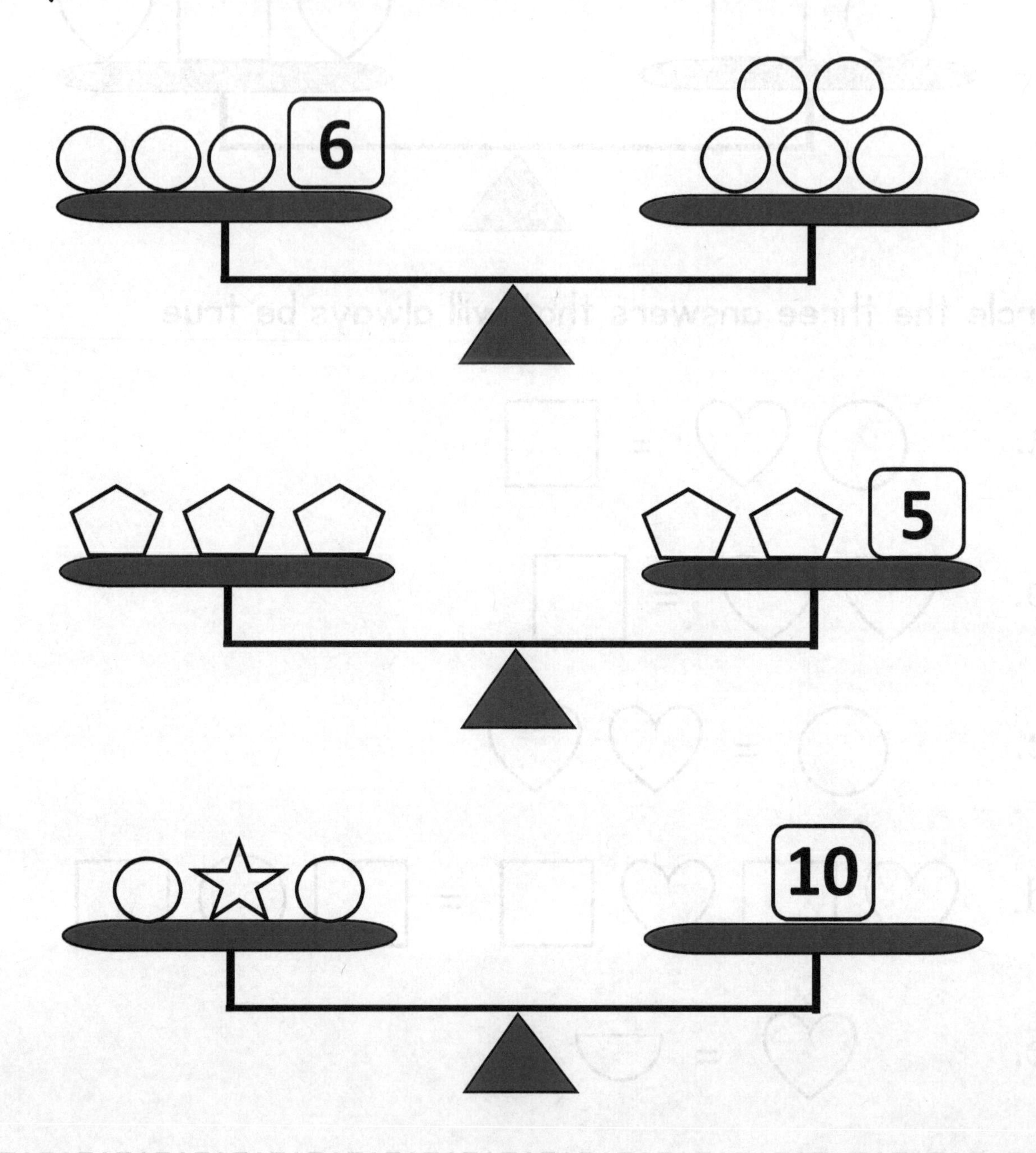

○ = ?

☆ = ?

⬠ = ?

Measurement : Game and Challenge

Use your math skills to find the value of each "?".

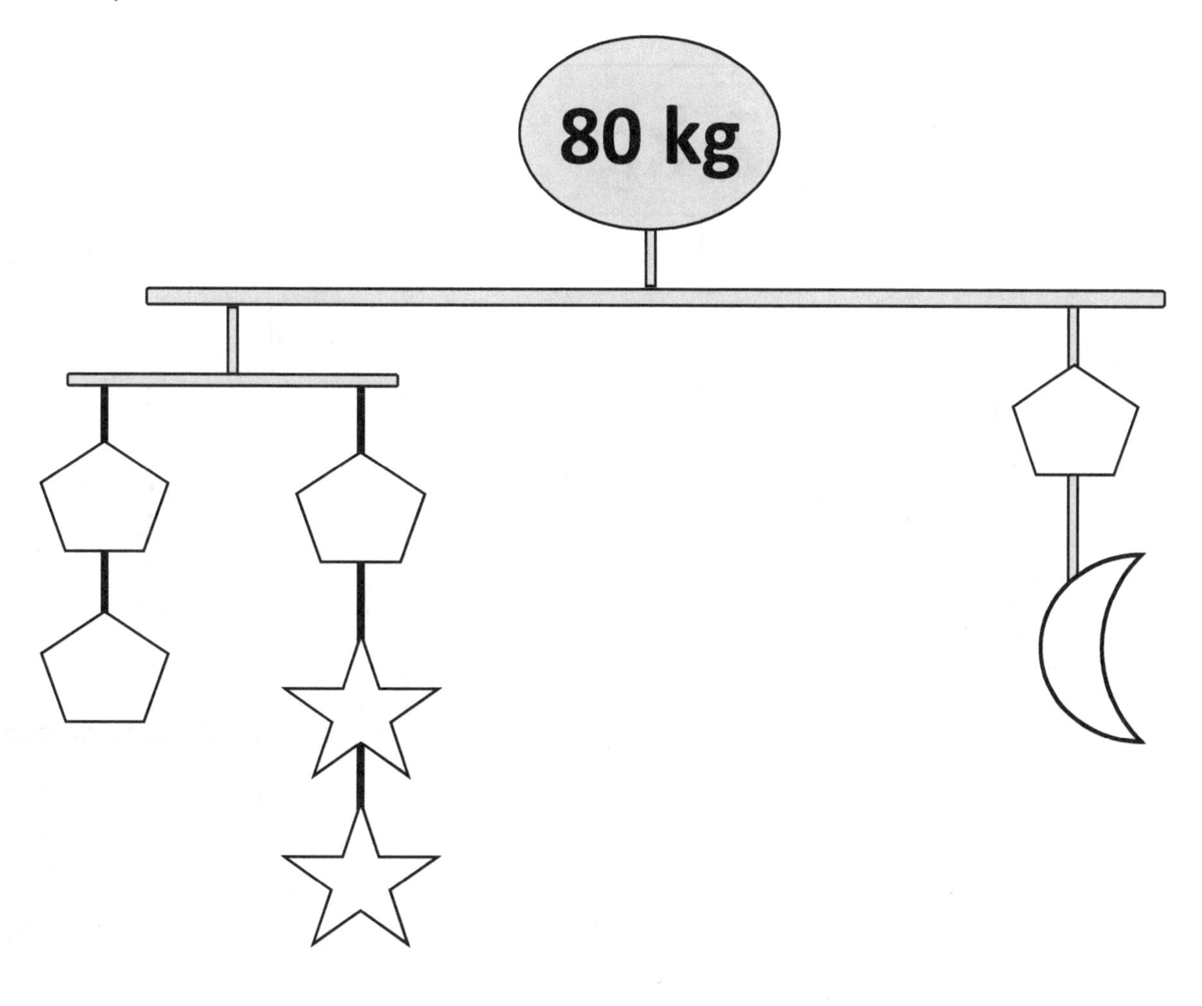

= ?

= ?

= ?

Measurement : Game and Challenge

Use your math skills to find the value of each "?".

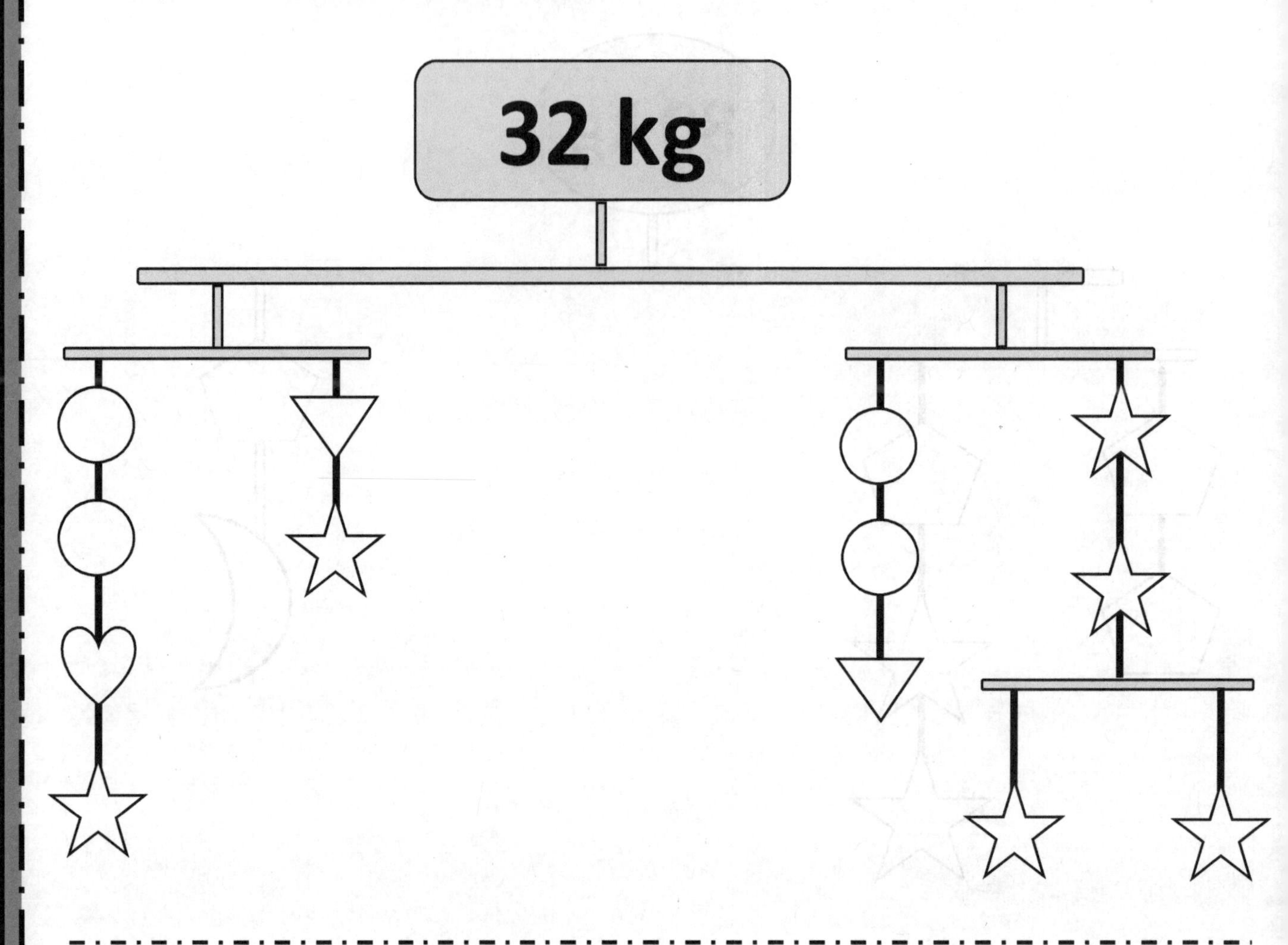

♡ = ?

○ = ?

☆ = ?

▽ = ?

Measurement : Game and Challenge

You are doing some gardening, and need exactly 4 liters of water to mix up some special formula for your award winning roses.

But you only have a 5-liter and a 3-liter bowl, but do have access to plenty of water.

How would you measure exactly 4 liters?

5-liter

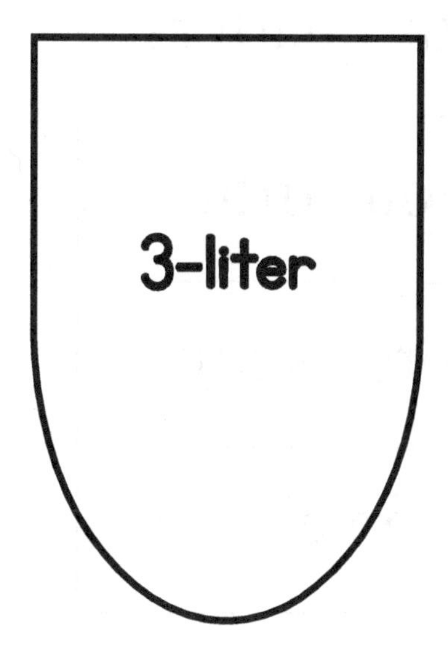

Fractions

Draw a line to match the fraction to the words.

One half	$\frac{2}{3}$
One third	$\frac{4}{5}$
One quarter	$\frac{1}{4}$
Two thirds	$\frac{8}{9}$
Three quarters	$\frac{1}{2}$
Four quarters	$\frac{3}{6}$
Four fifth	$\frac{1}{3}$
Three sixth	$\frac{4}{4}$
Seven eighth	$\frac{3}{4}$
Eight nineth	$\frac{7}{8}$

Fractions

Equal Parts

Write "Halves", "Thirds", "Quarters", "Sixths" or "Eighths" under each shape.

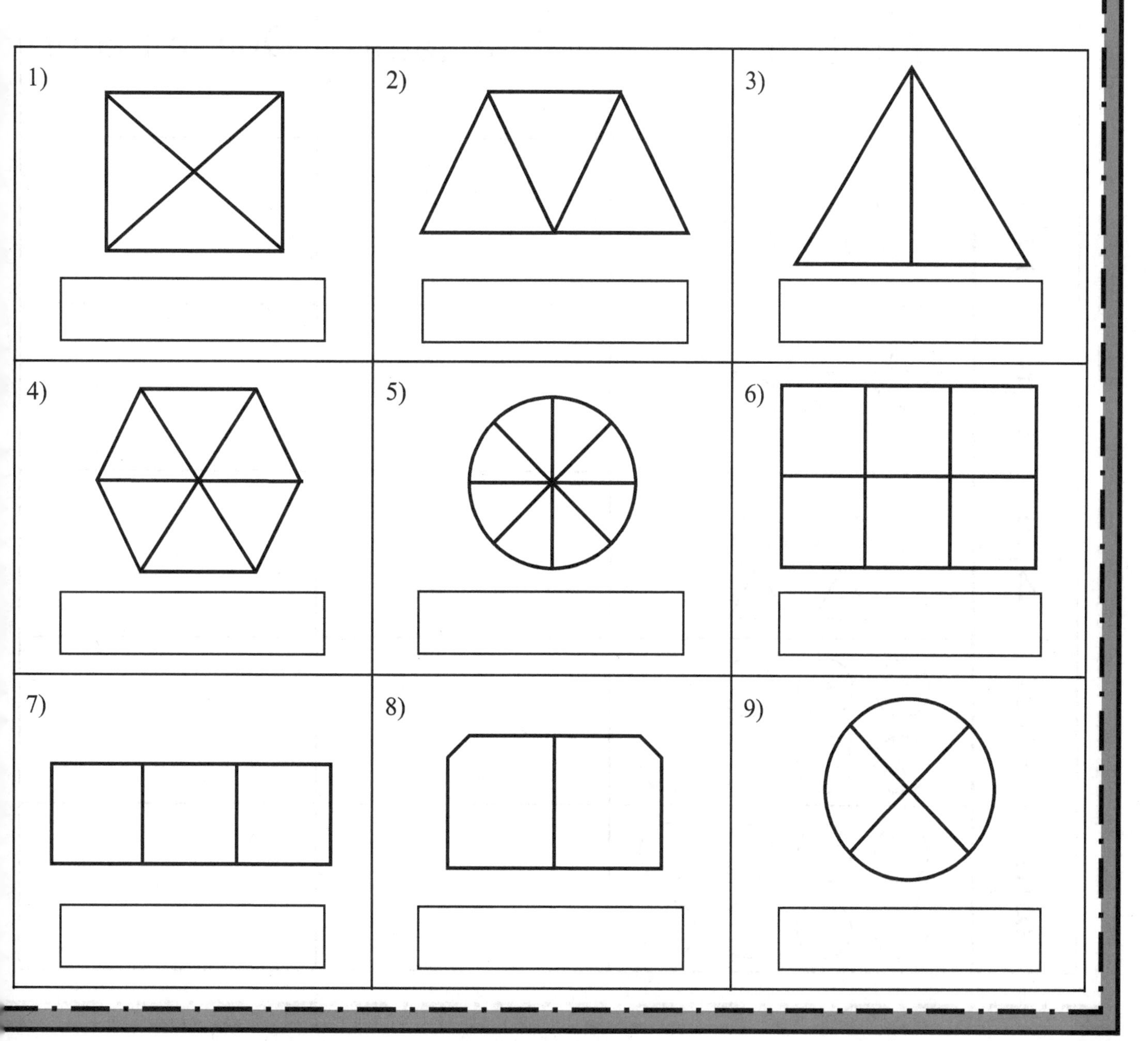

Fractions

Identify Numerators and Denominators

Fill in the table.

Fraction		Numerator	Denominator
$\frac{1}{2}$			
$\frac{2}{3}$			
$\frac{3}{4}$	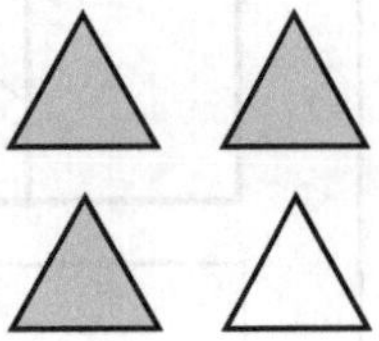		
$\frac{4}{7}$	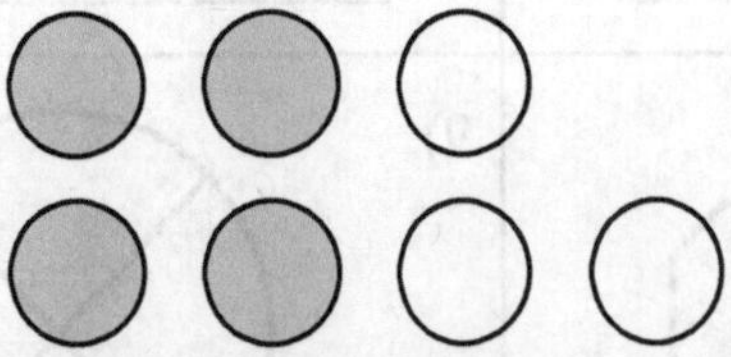		
$\frac{5}{8}$	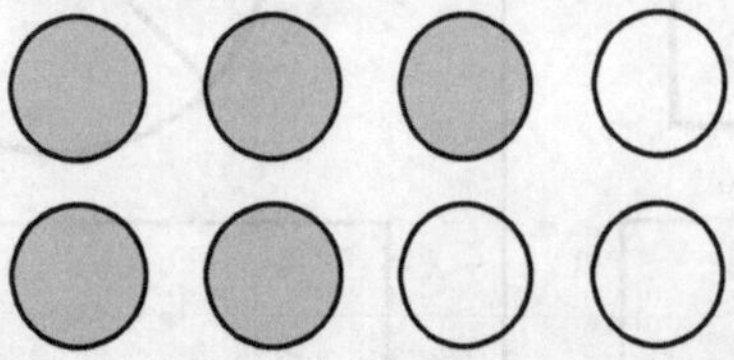		

Fractions

Identify Numerators and Denominators

Write the fractions in the first column.

Fraction		Numerator	Denominator
		3	5
		2	4
		1	6
	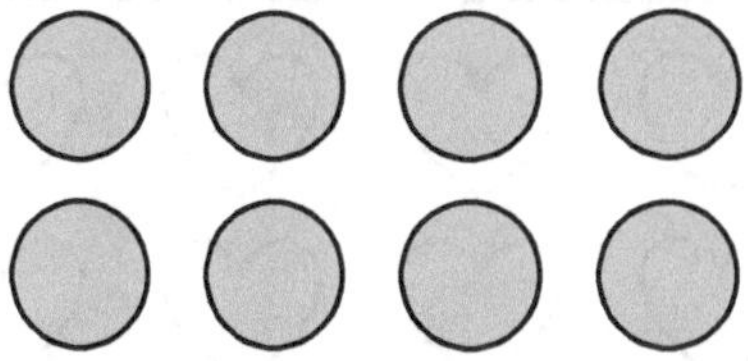	8	8
		6	9

Fractions

Coloring to make fractions

Color in the fraction shown of each shape.

	$\frac{2}{5}$		$\frac{2}{3}$
	$\frac{1}{4}$		$\frac{3}{3}$
	$\frac{6}{8}$		$\frac{4}{5}$
	$\frac{6}{7}$		$\frac{3}{6}$
	$\frac{2}{9}$		$\frac{7}{8}$
	$\frac{8}{10}$		$\frac{7}{7}$

Fractions

Writing Fractions

1) What fraction of the triangles are black?

2) What fraction of the stars are black?

3) What fraction of the circles are black?

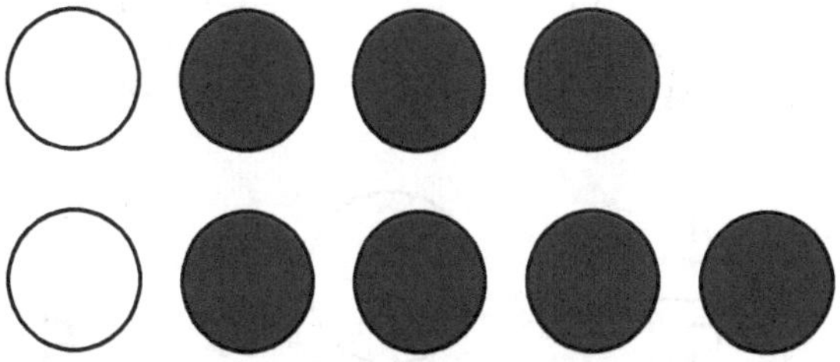

4) What fraction of the squares are black?

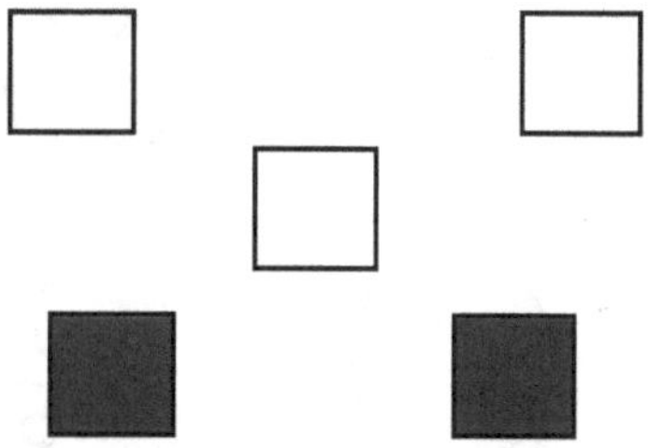

5) What fraction of the pentagons are black?

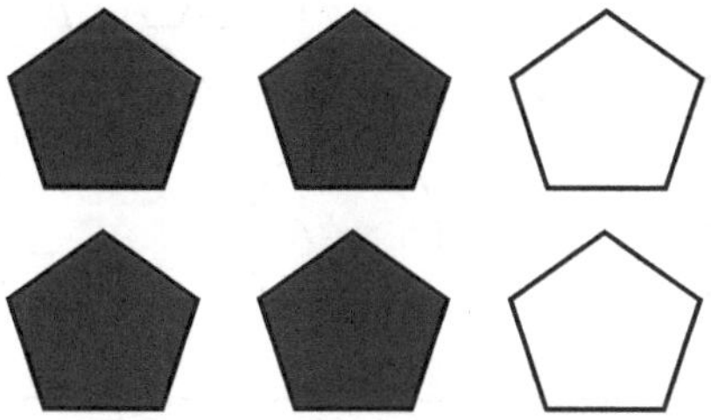

6) What fraction of the arrows point up?

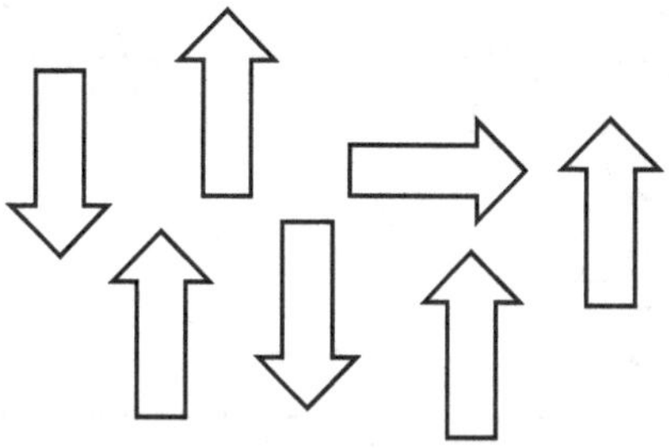

Comparing Fractions

Write the correct sign < or >.

$\frac{3}{6}$ ◯ $\frac{4}{6}$	$\frac{2}{4}$ ◯ $\frac{1}{4}$
$\frac{5}{8}$ ◯ $\frac{7}{8}$	$\frac{4}{5}$ ◯ $\frac{3}{5}$
$\frac{6}{9}$ ◯ $\frac{8}{9}$	$\frac{5}{7}$ ◯ $\frac{1}{7}$

Fractions

Comparing Fractions

1) What fraction of the vegetables are carrots? _________

2) What fraction of the vegetables are aubergine? _________

3) Which fraction is greater? _________

1) What fraction of the fruits are apples? _________

2) What fraction of the fruits are strawberries? _________

3) Which fraction is greater? _________

1) What fraction of the animals are rabbits? _________

2) What fraction of the animals are cats? _________

3) Which fraction is smaller? _________

Word Problems

Read and answer each question.

There are 12 buses in the parking lot. 3 buses are parked on the left and the other cars are parked on the right.

1. What fraction of the buses are parked on the right?

2. If 5 buses are black, what fraction of the buses are black?

3. If 3 buses are green, what fraction of the buses are green?

4. If $\frac{6}{12}$ of the buses are mini, how many mini are there?

5. During lunch time, $\frac{4}{12}$ of the buses left the parking lot. How many buses are left?

Fractions

Word Problems

Read and answer each question.

There are 22 students in a class. 9 of them are girls.

1. What fraction of the class are girls?

2. What fraction of the class are boys?

3. If 13 students ordered orange juice, what fraction of the class ordered orange juice?

4. If $\frac{6}{22}$ of students brings pizza for lunch, how many students have pizza for lunch?

5. $\frac{18}{22}$ of the students bring their parent consent forms for their field trip. How many of the students have not brought in their forms?

Fractions

Word Problems

Read and answer each question.

There are 27 books on a bookshelf. 13 of the books are large and the rest of them are small.

1. What fraction of the books are small?

2. If $\frac{23}{27}$ of the books are in English and the rest are in French, then how many French books are?

3. If $\frac{11}{27}$ of the fiction books are science fiction and $\frac{13}{27}$ are historical fiction, are there more science fiction or historical fiction?

4. There are only three kinds of fiction on the shelf: science fiction, history fiction and classics. How many classics are there?

5. If 2 of the science fictions are taken away, what is the fraction of the science fictions that is left behind?

Fractions

Word Problems

Read and answer each question.

James and Mark are visiting the pet store.

1. They see 18 rabbits. 6 of the rabbits are white and 12 of them are brouwn in color. What fraction of the rabbits are brouwn?

2. There are 8 squirrels on display. 5 of the squirrels are white and 3 of them are grey. What fraction of squirrels are grey?

3. There are 10 cats waiting to be adopted. 7 of them are kittens. What fraction of the cats are kittens?

4. There are 13 packs of pet food on the shelf. 4 of them are for squirrels and the rest of them are for rabbits. What fraction of the food is for rabbits?

5. The pet store has 8 people working there. Three eighths of the people are working in the grooming center. Four eighths of them are working at the cashier. The rest of them are helping customers in the store. How many staff members are helping customers?

ANSWER KEYS

The answers are only available to the word problems, challenges and games.

Answer Keys

Page 29

1) 19 cents 2) 14 marshmallows 3) 6 apples

Page 30

1) 20 flowers 2) 27 slices 3) 14 books

Page 31

1) 82 days 2) 49 exhibits 3) 40 inches

Page 32

1) 25 points 2) 40 pieces of candy 3) 80 CDs

Page 33

1) 12 biscuits 2) 2037 pages 3) 244 pounds.no.

Page 34

= 2

= 1 ;

= 1

= 2

Page 35

= 3

= 2

= 1

Page 36

= 7

= 5 = 8

= 4

Answer Keys

Page 37

 = 8 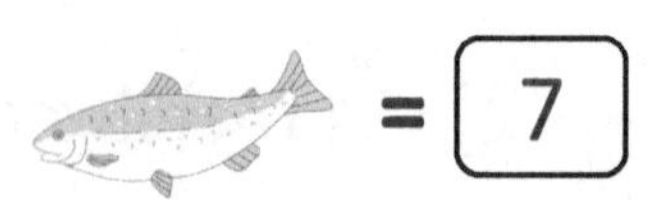 = 7  = 15

Page 38

 = 8 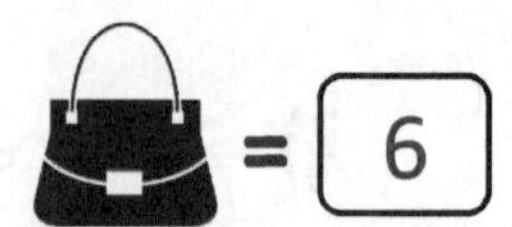 = 6 = 4

Page 39

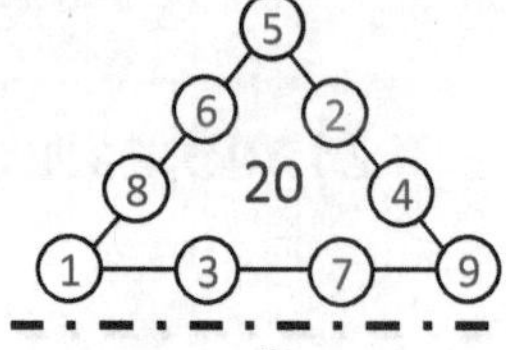

Page 40

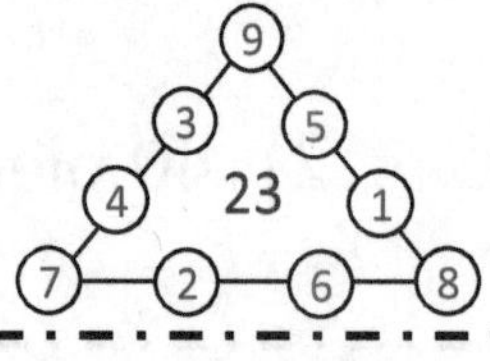

Page 48

1) 575 points 2) $466 3) 19 inches

Page 49

1) 22 kicks 2) 73 blocks 3) 9 whiteboards and 3 boxes.

Page 50

1) 2 cars 2) 54 glasses 3) 15 green triangles

Page 51

1) 5 points 2) 15 students 3) 43 points

Answer Keys

Page 54

1) 15 light bulbs 2) 38 patients 3) 57 cars

Page 55

1) 54 pencils 2) 47 cookies 3) 11 lockers

Page 56

1) 22 points 2) 16 points

Page 57

1) 28 passengers 2) 16 passengers

Page 58

 = 12 = 4 ;  = 10 = 20

Page 59

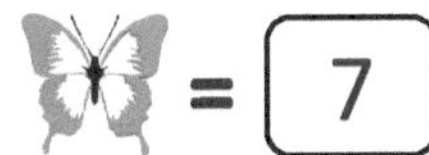 = 7 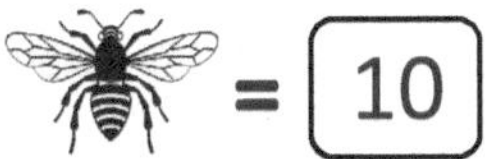 = 10  = 5

Page 60

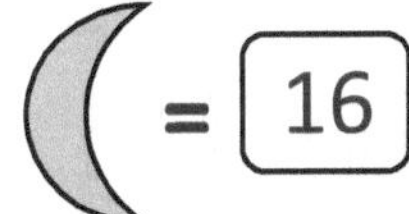 = 16 = 11 = 14 = 7

Page 61

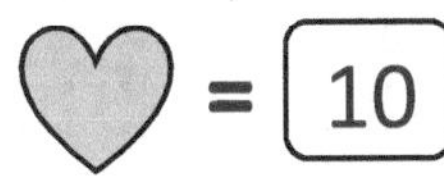 = 10 = 6 = 20 = 17

Answer Keys

Page 62

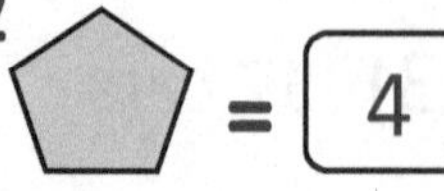

= 4 = 13 = 14 = 8

Page 63

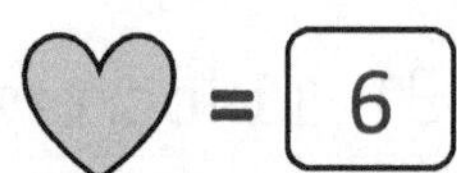

= 6 = 20 = 7 = 5

Page 64

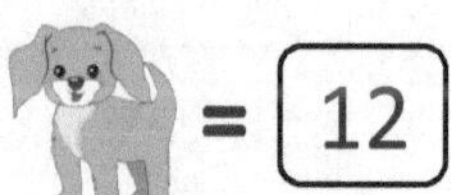

= 12 = 8 = 4 = 7

Page 65

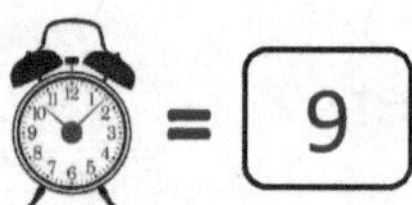

= 9 = 6 = 16 = 11

Page 66

= 4 = 9

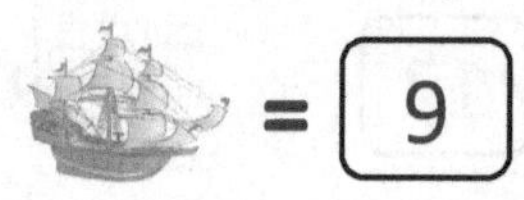

= 9 = 3

Page 79

1) 18 students 2) 12 pairs of pens 3) 12 sheets 4) 35 questions
5) 32 stickers

Page 80

1) 24 children 2) 18 pets 3) 24 letters 4) 18 guests 5) 16 cars

Page 81

1) 8 sheets 2) 25 muffins 3) 16 cups 4) 18 eggs 5) 20 cubes

Answer Keys

Page 82

 = 2   = 3 ;  = 1    = 6

Page 83

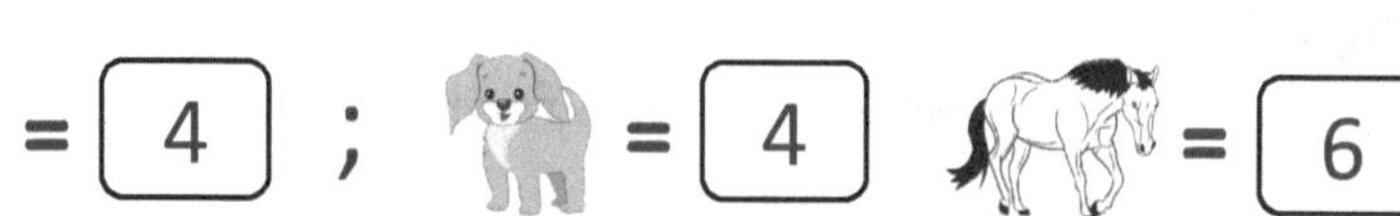

 = 3    = 4 ;  = 4   = 6

Page 101

1) 10 and a half hours 2) 3 o'clock in the afternoon 3) 12:30 in the afternoon 4) two hours 5) 8 o'clock in the evening

Page 102

1) 3 and a half hour 2) 9:15 in the evening 3) 45 minutes later than scheduled 4) Half past eight in the evening 5) five to ten

Page 114

 21

Page 115

c. d. e.

Page 116

 = 3    = 4    = 5

Page 117

 = 30   = 5   = 10

Answer Keys

Page 118

♥ = 4 ★ = 2 ▼ = 6 ● = 1

Page 119

Fill the 5-liter bowl. Then fill the 3-liter bowl from the 5-liter bowl. You will now have 2 liters left in the 5 liter bowl.

Empty the 3-liter bowl, and then transfer the 2 liters from the 5-liter bowl into it.

Now fill the 5-liter bowl again, then pour water carefully from the 5-liter bowl into the 3-liter bowl until it is full - exactly one more liter.

The 5-liter bowl now has exactly 4 liters.

Page 128

1) 9/12 2) 5/12 3) 3/12 4) 6 buses 5) 8 buses

Page 129

1) 9/22 2) 13/22 3) 13/22 4) 6 students 5) 4 students

Page 130

1) 14/27 2) 4 French books 3) 13/27 > 11/27 4) 3 classics books
5) 9/27

Page 131

1) 12/18 2) 3/8 3) 7/10 4) 9/13 5) One person

Made in United States
Troutdale, OR
12/04/2024